Togaev Isamiddin Safarovich

STRUCTURALLY-DECRYPTABLE COMPLEXES OF SOUTH NURATAU AND THE EFFECTIVENESS OF THEIR USE IN REGIONAL GEOLOGICAL STUDIES

Monograph

ISBN 978-93-6908-863-8

Book	:	Structurally-Decryptable complexes of South Nuratau and the effectiveness of their use in regional Geological Studies
Author	:	Togaev Isamiddin Safarovich
Publisher	:	Taemeer Publications
Year	:	'2025
Pages	:	90
Title Design	:	*Taemeer Web Design*

Togaev I.S.
Structurally-decryptable complexes of South Nuratau and the effectiveness of their use in regional geological studies. Monograph. Togaev I.S.

This monograph focuses on addressing practical geological issues and studying structural-decoding complexes using remote sensing materials, serving as one of the advanced modern technologies that ensure high economic efficiency in geological exploration and research works.

The comprehensive analysis of cosmogeological and geological-geophysical data in the Southern Nurata region allows the identification of prospective areas based on a new type of cosmogeological map-remote foundation. These methods are particularly significant for the high reliability of identifying prospective areas through visual and automatic decoding methods (with a field observation phase) of geological and cosmogeological research. This, in turn, forms the basis for future scientific research in this direction.

The method of identifying prospective areas based on remote foundations, developed as a result of decoding digital space images for the Southern Nurata region, allows for its application in production practices, enhancing the efficiency of geological exploration works.

This monograph is intended for specialists using geoinformation systems in geological mapping, the search and exploration of mineral deposits, as well as for university students and researchers in the field of geology.

Editor-in-Chief:

M.S.Karabaev – Doctor of geology and mineralogy sciences, professor.

reviewers:

Kh.D. Ishbaev – Doctor of geology and mineralogy sciences, professor;
I.B.Turamuratov – Doctor of philosophy (PhD)

INTRODUCTION

Extensive use of remote sensing data to improve the efficiency of geological exploration shows that it is crucial in identifying mineralized areas and promising structures at low costs. Consequently, numerous global measures are being implemented to widely apply remote sensing methods. The benefit of remote sensing in geological exploration lies in the comprehensive interpretation of existing geological data with digital space images using geoinformation technologies, which is particularly important for identifying new promising areas.

Currently, many scientific studies are being conducted in developed countries to use digital multispectral space images for identifying promising areas. These studies focus on obtaining new data and highlighting highly prospective areas for mineralization and favorable structural positions by reprocessing the physical-geological characteristics of known reference objects in each region using digital space images. This approach allows for a more thorough justification of the types of mineralization formation in geological structures.

In our country, considerable attention is being given to the application of modern remote sensing methods in geological exploration activities, and significant progress has been made. For example, a new innovative four-stage technological scheme for observation and automatic decoding has been developed. Conducting scientific research focused on identifying promising areas based on the reprocessing of digital space images for the Southern Nurata region is seen as important in this context.

It is worth noting that although the Southern Nurata region has been adequately studied geologically, geophysically, and geochemically, some unresolved issues remain. Additionally, in geological research, the study of the geology, tectonics, and patterns of mineral deposit locations in the Southern Nurata region, as well as the decoding and interpretation of remote sensing materials, has been reflected in the scientific research of many scholars. These scholars include O.M. Borisov, Sh.E. Ergashov, A.K. Glukh, E.B. Bertman, Yu.I. Loshkin, A.K. Bukharin, Ya.G. Katz, S.S. Schultz, V.S. Korsakov, N.T. Kochneva, I.N. Thomson, P.F. Ivankin, F.A. Usmanov, Z.M. Abduazimova, A.K. Nurkhojaev, Kh. Khodzhibekov, I.A. Pyanyovskaya, M.K. Turapov, Yu.V. Shumakov, Kh.A. Akbarov, R.Kh. Mirkamalov, Kh.A. Toychiev, S.M. Koloskova, N.I. Nazarova, A.R. Asadov, R.S. Khan, A.A. Abdurakhmanov, O.T. Zakirov, A.R. Avezov, E.R. Khachaturyan, I.G. Potorzhinsky, D.A. Magzumova, S.T. Maripova, P. Kronberg, T. Graupner, R. Seltmann, H. Boorder, U. Kempe, and many others.

Considering that much of the studied region is inaccessible, the comprehensive use of remote sensing materials enhances the reliability of new geological data. It provides opportunities to study the distribution of many mineral deposits and promising structures.

Modern cosmogeological research methods, compared to traditional field geological methods, require relatively less time and expense to perform large-scale geological fieldwork. Additionally, the use of remote sensing data reveals indicators of the spatial distribution of intrusive formations that control the formation of mineralized structures, indicating potential areas for promising structures and mineral deposits. Moreover, the identification of ring structures and structurally distinct structural-decoding complexes, along with their associated potential promising area boundaries, becomes feasible.

Integrating cosmogeological materials with geological and geophysical data provides additional information, which is crucial for solving regional geological problems. This combination enhances the practical significance of the study by offering a comprehensive approach to identifying and analyzing promising areas and structures.

The research employs modern computer tools to visually and automatically decode space images. This includes separating structural-decoding complexes and classifying linear and ring structures visible in space images. Traditional geological exploration methods, such as lithological-structural sections, sample collection, observation points, point approbations, and laboratory analysis results, are integrated into the study.

The results identified structural-decoding complexes that exhibit geological features in high-resolution space images. Geological decoding signs of tectonic structures and areas covered by Meso-Cenozoic deposits were also identified. The method for creating large-scale cosmogeological maps based on remote sensing materials has been refined. Scientific foundations for identifying promising areas in the Southern Nurata region have been developed, and the optimal channel ratios for automatic decoding of digital space images using the RGB system (Red, Green, Blue) have been determined.

The practical significance of these results lies in their application to production practices. The new method for identifying promising areas based on the decoding of digital space images enhances the efficiency of geological exploration in the Southern Nurata region.

The author extends deep gratitude to Dr. A.Q. Nurkhodjaev, Doctor of Geological and Mineralogical Sciences, for his valuable insights and ideas that contributed to the results presented in this monograph. Additionally, the author thanks everyone who provided advice and assistance during the preparation of this monograph. We hope that this book will be beneficial for a wide range of geological readers

I.S. Togaev

CHAPTER I. THE GEOLOGICAL STUDY STATUS OF THE SOUTH NURATA REGION

1.1. Brief history of geological and cosmogeological research

The mineral wealth of the Southern Nurota area has been known for a long time, and the names of the places (Oltinkozgan, Oltinsoy, etc.) and the existing signs of gold mining, that is, the preserved remains, clearly testify (Riskidinov et al., 2012). Since the end of the 19th century, geological research has been carried out on a large scale in this area and adjacent areas. M.V. Mushketov, who conducted research in 1878-1879 and 1886, G.D. Romanovsky noted that the rocks of the Silurian period are widespread in the area (Pyanovskaya, 1984). 30 years later, under the leadership of V.M. Nikolaev, he began the planned geological survey of the Southern Nurota area (V.A. Nikolaev, 1925).

In 1927-1928, the Langar mine was opened by S. V. Kultiasov's party, and during the research work, B. A. Nikolaev's stratigraphic scheme was clarified (Pyanovskaya, 1984, Khan, 1989, 2002).

In 1932-1935, N.A. Smirnov, B.I. Markelov carried out small-scale geological mapping, as a result of which they created a geological map of the entire Nurota region.

In 1935-1937, under the leadership of K. L. Boboev, metamorphic complexes and their lithological-petrographic characteristics were thoroughly studied in the western part of Southern Nurota (Pyanovskaya, 1984, Riskidinov 2012).

In 1938, B.A. Grozin found the old mining facilities where gold was mined in Altynkozgan and M.D. Troyanov carried out detailed prospecting until 1959. The gold mine of Altynkozgan was opened.

In 1939, the geological description of the Karatov Mountains was widely covered by T.V. Zvenko. In 1946-1950, K.B. Shulyatnikov, I.M. Shamshurin studied the contact aureoles of the Karatov intrusive and identified the deposits with an industrial concentration of tin (Suluvkyz et al.).

As a result of 1:100,000-scale geological mapping by V. M. Zheleznov in 1952, strongly metamorphosed rocks (marbles and crystallized shales) belonged to the lower Paleozoic, and Silurian deposits were divided into the Lower Llandover, Upper Llandover, Wenlock and Ludlow layers (Khan, 1989, 2002).

In 1955-1959, H.V. Riskinov, G.S. Chikrizov created a new scheme of stratigraphy of Paleozoic deposits as a result of carrying out geological imaging-research works on a scale of 1:100,000 in the Southern Nurota area. Conducting small-scale geological imaging and research work in the Southern Nurota ridges has formed different views on the geological structure. Some researchers believed that the Oktov marbles lying in the core of the anticlinal fold are consistent with

Cambrian-early Paleozoic crystalline shales and Silurian deposits. Another group, including N.A. Smirnov, noted that the marbles of the Oktov series developed at the base of a synclinal fold, the wings of which consist of variously metamorphosed carbonate sediments (Khan, 1989, 2002, Riskidinov, 2012). H.M. Abdullaev, I.H. Hamroboev, I.M. Isamuhamedov, B.D. Vasilevskiy, N.A. Yakushev, N.D. Zlenko also mentioned different points of view on the formation of granitic intrusives. N.V. Nechelyustov, B.A. Baskin, V.A. Talalov, Yu.A. Chernyavsky believed that the intrusive was formed with the participation of one-phase assimilation processes (Khan, 1989, 2002). Based on the research results of M.M. Posokhova, H.V. Riskinova, G.S. Chikrizov, they came to the opposite conclusion about crystalline shales. According to them, the distribution of shales is closely related to intrusive bodies. Since the beginning of the study of the stratigraphy of the Southern Nurota ridges, problems regarding the age of the Oktov and Bakhiltov marbles have been addressed by V.A. Nikolaev (1925), B.F. Popov (1953), V.M. Zheleznov (1952) and other experts, according to whom the age of the marbles belongs to the Lower Paleozoic. According to N.A. Smirnov (1933), N.D. Zlenko (1950), G.S. Chikrizov (1952), H.V. Riskina (1958), K.K. Pyatkov (1961), Devonian-Carboniferous they believe that Based on Yu.A. Likhachev's data and comprehensive analysis of previous data, he concluded that the Oktov marbles and crystalline shales lying under it were formed in the Ordovician-Silurian period (Khan, 2002).

There are different points of view about the genesis of the igneous rocks of the Southern Nurota area. In early work, there were different opinions about the multiphase nature of the main intrusives (V.A. Talalov, Yu.A. Chernyavsky and others, 1958).

According to H.M.Abdullaev, I.M.Isamuhamedov, Oktov, Zarkainar batholiths were formed during the Hertzian stage of magmatism. I.H. Hamroboev developed a complete scheme of magmatism development and distinguished four different age Upper Silurian, Devonian, Lower-Middle Carboniferous, Upper Carboniferous complexes. The Oktov intrusive was included in the last complex (Loshkin, Tillyaev et al. 1965). According to I.H. Hamroboev, the granitic massifs of Nurota were formed in S2-3 (Khan, 2002).

Since 1961, geological imaging works on a scale of 1:50,000 have been carried out in South Nurota. As a result of 1:50000 scale imaging, Yu.I.Loshkin created a completely different new stratigraphic scheme.

H.V. Riskin (1966-1969), V.M. Heifits (1967), M.M. Posohovalar (1967) 1:100000 scale geological imaging data and new observations of the Southern Nurota Ranges and adjacent areas 1: 200,000 scale geological map published. P. T. Azimov conducted a number of works on the study of intrusive derivatives.

In 1965-1969, D.M. Ogaryov carried out geological imaging works on the scale of 1:50000 in the southern branch of the western part of the Southern Nurota ridges (Khan, 1989, 2002, Togaev, 2018).

In 1967, G.D. Shmulevich compiled the metallogenic prediction map of Western and Southern Uzbekistan by summarizing large-scale data. In these years, it was reflected in the monograph published by K.K.Pyatkov, I.A.Pyanovskaya, A.K.Bukharin.

In the 1970s, Y.V.Shumakov, D.M.Ogaryov continued the geological imaging works on the scale of 1:50000 in the northern and southern branches of the western part of the Southern Nurota Ridge. Based on the study of graptolites by Z.M. Abduazimova (1969), A.N. Golikov (1970), biostratigraphic zones were distinguished for the territory of Western Uzbekistan, including the Southern Nurota ridges (Khan, 1989).

In 1974, I.A. Pyanovskaya presented all the stratigraphic and magmatic units in the table of conventional signs for the 1:50000 scale geological maps of Western Uzbekistan. In 1979, I.A. Pyanovskaya made a geological map of the western part of the Southern Nurota region on a scale of 1:50000 (Pyanovskaya, 1984, Khan, 1989, 2002).

In 1975, 5 complexes and the last complex of granite dykes were separated in the Oktov batholith by E.P. Ikhozlar. In 1975, Z.A. Yudalevich combined 3 sub-complexes such as Darasay, Shorokh and Gatchin and distinguished only one Shorokh complex in the Oktov intrusive.

In the 1980s, in the territory of Western Uzbekistan, including Southern Nurota, the geological re-survey of previously imaged areas on a scale of 1:50,000 (GDP-1, GDP-2), aerogeological mapping on a scale of 1:50,000, deep geological mapping, and volumetric mapping of large gold deposits began.

In 1981, I.A. Pyanovskaya and others carried out a geological resurvey (GDP-1) on a scale of 1:50,000 in the Karakchatov Mountains without general prospecting.

In 1983, A.A. Pokrovsky distinguished two types of metamorphism with andalusite-sillimanite and kyanite-sillimanite based on the study of the metamorphism of the Southern Nurota Ridge, and each type of regional metamorphism was divided into facies and subfacies. The results were reflected in the monograph published by A.V. Pokrovsky in 1988.

In 1984, Y.V. Shumakov and others carried out geological re-survey and deep geological mapping in the eastern part of Southern Nurota and Gobduntov.

In 1989, a geological re-examination of the fold complex was carried out in the central part of the South Nurota Range by RS Khan et al. As a result, the existing stratigraphic scheme was fundamentally updated, the presence of shear tectonics, the boundary of metamorphic rocks and stratigraphic divisions were proved to be

8

inconsistent in space. The results of regional work in Northern and Southern Nurota in the 90s are described in the works of experts such as A.K. Bukharin, R.R. Usmonov and R.S. Khan.

In 1993, V.S.Korsakov and others made conventional signs on geological maps on a scale of 1:50,000 and are still the basis for mapping the territory of Western Uzbekistan (R.S.Khan et al. 2006). R.N.Abdullaev and Z.M.Abduazimova highlighted the main importance of paleontological methods in the reconstruction of the history of geological development of Southern Tien-Shan (Khan, 1989, 2002).

In 2000, in the works of A.K. Bukharin, a large number of structural-material complexes superimposed on each other were separated. When separating them, some structural-material complexes combined only one or two suites. The results of the work on the tectonics of the region are reflected in the scientific works of A.K. Bukharin, A.K. Pyatkov and R.R. Usmanov (2000). Based on the research work of E. B. Bertman, mineralization types corresponding to four ore formations were distinguished in the Nurota area, and it was noted that they play a major role in the research of gold ore fields.

In 2016, G.S. Abdullaev, F.G. Dolgopolov showed that the formation of tectonic structures observed in the oil and gas regions of Uzbekistan mainly reflects the nature of three external pressure forces (movement of lithospheric plates) of the modern (Neogene-Quaternary) Central Asian lithosphere (Abdullaev, Dolgopolov , 2016).

A brief history of cosmogeological research. The role of digital space photographs is of particular importance in studying the structure of the Earth's surface, the relationship between its forms and tectonic elements. The widespread use of digital space images from satellite systems in geological exploration at various scales has led to significant changes and innovations in the study of geological science. The study of remote sensing of the Earth began in 1783 with data from the first aerostat launched by the Mangolf brothers.

The first photo was taken in 1855. The use of a balloon photo of Paris for geological purposes dates back to 1858-1882. The first geological research was started by the French geologist Emme Tsivile from aerial photographs of the Alps (Gubin, 2004).

In the early years of the 19th century, the use of aerial photographs in the study of natural resources in the territories of the Russian state was widely introduced. On the initiative of Russian geologist academician A.E. Fersman, the first aerial photography institute was established in Leningrad (St. Petersburg). Since 1938, aerial photography has been widely used in geological imaging (Gubin, 2004).

In the second half of the 19th century, the interest in space research and the creation of technical means and the launch of the first astronaut into space laid the foundation for a new phase of remote Earth exploration. At the same time, the creation and launch of satellite systems in countries such as France in 1965, Japan in 1970, England in 1971, and India in 1980 led to changes in space exploration.

The Republic of Uzbekistan is also entering the world space research system with rapid images. The decree of the President of the Republic of Uzbekistan on February 12, 2018 "On space research and technology development activities in the Republic of Uzbekistan" gave a special impetus to the development of the field.

Almost 60 years since the launch of the first Earth satellite, new information has been gained in the study of the structure of the Earth's surface, the evolution of structures, and the study of natural resources.

On the basis of this information, a new method of geological research was created, that is, the method of aerospace research, which is now called the method of remote sensing of the Earth. As a result of the data of remote sensing of the earth, new achievements were made in solving some issues of geological structure, tectonics, structure of mining areas, neotectonics and other geology. Since 1956, aerial photographs have been used in some areas of Central Asia.

Since 1967, pilot-methodological works of small- and medium-scale aerial photogeological mapping have been carried out in mountainous areas. For the first time, ring-shaped structures were detected in several areas of Central Kyzylkum in space photographs (O.M. Borisov, A.K. Glukh, etc. 1982). Since 1978, geological organizations and scientific research institutes have been widely using space materials in geological imaging, prospecting, hydrogeological and other directions.

As a result of cosmogeological research, "Album of lineament and ring-shaped structures of Central Asia" (1:1 000 000, H.T. Tulyaganov, 1979), "Drawing of the main linear and ring-shaped structures of Kazakhstan and Central Asia" (1:1 500 000, A .K.Glukh., V.I.Vnuchkov) was created.

The second half of the 70s of the 20th century was considered the period of active use of aerospace methods in geology. Maps of aerophotogeological, space-photostructures on a scale of 1:500000 and 1:1000000 were created. The special information received for the study of the earth's natural resources through satellites laid the foundation for huge scientific and applied research.

In 1980-1990, new changes in the use of aerospace materials for geological purposes were reached. At this time, digital images of space data appeared, and their decoding was made possible using computer technologies. Improvement of the theoretical methodological developments of geological research based on the data of remote sensing of the earth, the solution of some problems related to the study was clarified. In this regard, B.V. Vinogradov, Y.G. Kats, Y.A. Kosygin, V.V. Kozlov, D.M. Trafimov, O.M. Borisov, A.K. Glukh, F.A. Usmanov, Sh.E. Ergashev, M.Kh. Khodzhibekov, A.Q. Nurkhodzhaev, Y.I. Loshkin, M.N. Thay, S.S.Shults, A.R. Asadov, Y.B. Aysanov, O. The scientific and methodological literature created as a result of the research work of specialists such as T.Zokirov, E.A.Agbalyants, N.Kh.Starostina, I.A.Pyanovskaya made a significant contribution to the field of remote sensing of the Earth. The study of the Earth's structure and natural resources based on remote sensing materials in terms of space technology and technology: has been rapidly increasing with images in developed countries such as the USA, France, India, Japan, Germany, Russia, China, Turkey and South Korea. Azerbaijan, Kazakhstan, including the Republic of Uzbekistan can be added to these countries. Analysis of the published literature shows that remote sensing of the Earth is still developing widely today. Large-scale cosmogeological maps of the Central Kyzylqum, Chotkal-Kurama, Nurota low-mountainous and other regions of the Republic of Uzbekistan embodying different landscape conditions were created.

At the same time, experiments were carried out on the use of Earth remote sensing materials to study endogenous mineralization and new photomesons in mining areas, optical photogrammetric properties of various rocks (Sh.E.Ergashev et al., 2001, 2009).

At the beginning of the 21st century, the results of processing materials obtained from digital space images of satellites and new methods at enterprises under the State Geological Committee of the Republic of Uzbekistan prompted the creation and interpretation of new types of cosmogeological maps at various scales. Digital images in cosmogeological research have raised remote methods to a new level in geological research. reflected in his scientific works. The relevance of this issue is determined by the decision of the President of the Republic of Uzbekistan dated January 17, 2007 No. 568 "On measures to fundamentally improve the organization of geological research and activities of the State Geological and Mineral Resources Committee of the Republic of Uzbekistan" became the reason for the creation of the center of sensing and GAT-technologies. Cosmogeological researches were launched on large, medium and small scales in areas with high geological potential (Nurkhodjaev, 2017).

Currently, tracking decoding of Landsat 7, Aster (TERRA), Quick Bird space images and SRTM radar images and automatic decoding based on ERDAS Imagine and ENVI computer programs are carried out. As a result, a new technological scheme for decoding space images was developed, structural-decoding complexes were separated in geological structures, and a new type of cosmogeological map- remote basis was created (Nurkhodjaev et al., 2016, 2017, 2018, 2019).

1.2. Overview of Geophysical and Geochemical Studies

Large-scale gravimetric and magnetometric mapping works have been carried out in almost all mountainous areas of Southern Nurota. Seismic profiling works were mainly carried out in areas where Mesozoic and Cenozoic deposits are developed on the plains. Different types of electro-exploration, magnetic prospecting and other types of geophysical research were carried out in areas with specific ore content. Recently, detailed geophysical analyzes were carried out in separate prospective sections (Koloskova, 2005).

In 1955, Sh.A. Chembarisov, P.V. Khramishkin carried out 1:25,000 and 1:50,000 scale magnetic prospecting and metallometric mapping works on the Karatov and Oktov intrusives.

In 1958, V.M. Fomina, Yu.S. Shmanenko carried out 1:200,000-scale graphic search mapping in the Southern Nurota region. In 1964, under the guidance of G.A. Ivanov, the depths of the Paleozoic foundation were determined as a result of seismic exploration in the valley of the Sanzor river in the folds of the Nurota mountain ranges, and granitoid massifs and faults were revealed. In 1966, a schematic geological structural-tectonic map on a scale of 1:100,000 was made. Prospects of the Kara-Machit and Quray gold occurrences were expanded based on 1:10000-1:2000 scale electrical prospecting (TP) and geochemical mapping data conducted by N.G. Degtyarev and others in 1967-1969. In 1968-1970, on the basis of G.N. Korobeynikov and others, a comprehensive geological-geophysical map was created in the area of the Altynkozgan mine field, which includes gold, margium, tungsten, tin, lead, beryllium and other metals. Three high fracture zones accompanying with. One of the methods used under the leadership of Y.F. Dyukov in the search and complex geophysical works in the areas of Kholbashi, Mavlon-Beshbuloq, Sarmich-Oltinkozgan is lithogeochemical mapping on a scale of 1:25000. Prospective anomalies of gold, silver and other metals were identified, which were recommended for prospecting. Selected samples were analyzed for elements within a certain range. In the process of analysis, a limited amount of seeding method was used (Koloskova, 2005, Khan, 1989, 2002).

As a result of summarizing geophysical and other data, G. N. Korobeynikov and others identified three zones of sulphide mineralization in the Sarmich-Oltinkozgan mining area in 1976. In 1972, as a result of aerogamma spectrometric mapping on a scale of 1:25000 by A.V. Evstigneev and others, promising areas were allocated by N.G. Degtyarev and others.

In 1972, Y.F. Dyukov and others determined that the positive magnetic anomalies correspond to the distribution of tectonized limestone-shale deposits of the Jivachisoi suite, where the Arab and II-Sarmich gold manifestations are located, and

12

the linear stockworks of Altinkozgan and Berkut belong to the zones of tectonic disturbances recorded in positive magnetic fields. based on.

In 1973, L.N. Kotlyarev carried out a 1:25,000 scale aeromagnetic mapping in the Southern Nurota region, and as a result of these works, the anomalies confirmed in the magnetic mapping of the earth's surface were shown in many respects. In 1974, G.A.Ivanov and others carried out generalization taking into account the existing geological results. As a result, the maps of the magnetic field and specific gravity anomalies did not record the Yengogli, Kholboshi and Karatov intrusions. In 1985, during the general search for non-ferrous metals, A.N. Kostyuk conducted magnetic mapping on a scale of 1:25000. In 1987-1991, A.N. Krivolapov carried out gravimetric mapping in the Southern Nurota region.

As a result, it was determined that the Karatov, Biran, Chettik, Polvontosh intrusives form the southern branches of the massif, and the Oktov, Bitob, Zarqaynar, Yong'ogli, and Kholbashi intrusives form the northern branches (R.S. Khan, 2006). On the map, deep earth cracks and promising sections typical of tungsten-molybdenum skarn are distinguished.

Geochemical works. These works were started in the early 1950s in the prospective plots in the Southern Nurota area on a scale of 1:25,000-1:100,000 (Riskidinov, Movlanov, 2012). In 1959, 1:500000 scale lithogeochemical imaging was carried out by E.G. Neusser. These works were carried out in the central part of the Southern Nurota Mountains, as a result of which geochemical anomalies of mercury, tin, tungsten, molybdenum, margium, copper, nickel, cobalt and others were revealed. The lower limits of the sensitivity of the analysis to the margin did not exceed 0.01%. The analyzes were carried out by the evaporation method.

In 1963, V. M. Tushov carried out lithogeochemical mapping works on all areas in the Southern Nurota region on a scale of 1:50,000. N.G. Degtyarev, L.V. Kvasovalar in 1967 re-(control) goldmetric analysis of samples of all previous works, in which simultaneous re-interpretation of spectrogram plates and lithogeochemical mapping on scales 1:25000-1:500000 were carried out. As a result of these works, the Marjonbulok, Sarmich, Subashi deposits were revealed, and the Karatov mineralized zone was identified (Riskidinov et al., 2012).

In 1965, lithogeochemical imaging works on the northern endo- and exocontacts of the Karatov intrusion were carried out by B.A. Gorshkov on a 200x20m grid at a scale of 1:25000. As a result of this mapping work, anomalies of niobium, tin, tungsten, lead and other elements were established (Khan, 1989). In 1969, N.G. Degtyarev, by summarizing previous materials, interpreting and reinterpreting spectrograms, defined spreading halos of margiumush, gold, silver, non-ferrous and rare metals. Prospective plots for gold were shown on the prepared prediction-metallogenic maps (Subashi-Sarmich-Oltinkozgan, etc.).

G. N. Korobeynikov and others carried out detailed inspection-evaluation and geophysical research in the Altynkozgan mining area. General geological-geophysical maps of the area on a scale of 1:25,000 were made based on the results of the work. Three high cleavage zones were distinguished in it, accompanied by halos of gold, margium, tungsten, tin, lead, beryllium and other metals (Koloskova, 2005). Since the 1970s, almost all geological mapping, prospecting and thematic work has been accompanied by lithogeochemical imaging at scales of 1:50,000 and larger. (Y.V. Shumakov, 1979, 1984; R.S. Khan, 1989, 1998; L.M. Kubrakova, 1984; E.B. Bertman, 1996; S.M. Koloskova, 2005). Based on their results, prospective gold mining areas have been allocated. Dozens of promising sites were recommended for prospecting, among which Bitob and Bashtut and other fields were given priority. As a result of geochemical work, a number of anomalies of tungsten, tin, mercury, lead and other minerals were revealed.

Summary. Although the South Nurota region has been well studied geologically, geophysically and geochemically, it was found that there are still some unresolved issues. As a result, the complex use of remote sensing materials in the Southern Nurota area made it possible to increase the level of reliability of geological data and study promising structures with new data; Research of natural resources on the basis of materials of remote sensing of the earth has gained special importance in their rational use and in solving tasks of applied geology in regional geological research works; The use of the results of geological research and the reanalysis of fund materials will serve to complete new geological information.

CHAPTER II. METHODS FOR DECODING REMOTE SENSING MATERIALS OF THE EARTH

Decoding space images means separating the natural and man-made sources reflected in the images. High-quality analysis of aerial photographs depends on the experience of the analyst, his ability to see (K. A. Toychiev, A. K. Nurkhodjaev et al., 1999). There are visual and automatic methods of decoding. We will dwell on them below. It is of particular importance to ensure the identification, opening and reading of the geological objects shown in space curats in terms of image color, color, shape and size, as well as their qualitative and quantitative indicators in their mutual proportions, directly or with the help of technical means (Nurkhodjaev et al., 2016, 2017, 2019, Togaev, 2018).

2.1. Preparation of Remote Sensing Materials of the Earth and Enhancement of Geological

Preparation of materials for remote sensing of the Earth is one of the main goals to be solved in space images, depending on the geological tasks. The preparation of space photos can be described as follows based on the defined geological tasks: - Selection of space photo materials for the research area of Earth remote sensing materials; - correcting the level of illumination for primary processing of space photo materials obtained from satellite systems, geographic linking of space photos and decoding in further work; - special processing to increase the informativeness of the data received from modern satellite systems and calculate the optimal parameters for subsequent computer and visual decoding; - identification of geological objects, preparation of the initial version of the cosmogeological drawings and remote basis of the studied scale and decoding (reading) of high-resolution images (Glux, 2009, Nurkhodjaev et al., 2016, 2017, 2019, Khachaturyan, Avezov, Glux, 2011, Togaev, 2018).

Preliminary processing of Earth remote sensing data is carried out using special tools, taking into account tasks, methods and developed algorithms for processing modern space images from digital satellite systems. Earth remote sensing data processing (image processing) means the process of performing work on space images of satellite systems in a prescribed manner, including their correction, modification and improvement.

Processing is carried out manually, instrumentally and automatically using computer technologies. Modern space image processing is versatile and includes noise filtering, geometric correction, local contrast enhancement, sharpening, image reconstruction, and more. Image processing methods are divided into two classes: frequency domain processing and spatial processing. Methods of processing space images in the frequency domain are based on the recommendations of T. Stockham,

Ch. Hall, E. Hall, which are compatible with human vision (Khachaturyan., Avezov, Glux, 2011). Processing in the frequency domain leads to complex calculations. Filtering is used in image processing not only to improve image quality, but also to identify objects in images (Glux, Mekhmonkhodjaev, Kim 2003). Defining a certain frame of the photo allows you to identify any object, for example, water, etc. Two labeling methods: The first method is to assign high luminance levels to the study area and low luminance values to all other areas, and the second method is to assign high luminance values to the region of interest without changing the background values of the image (Gonzalez, 2006). Increasing the geological information of Earth remote sensing materials first of all depends on the processing of early seasonal remote sensing images of the Earth's surface and the decoding of high-resolution space images. In this research work, satellite images from satellite systems (Landsat 7 TM, Aster, etc.) were extensively used in the preparation of earth remote sensing materials and geological data acquisition (Figure-2.1.1-2.1.2).

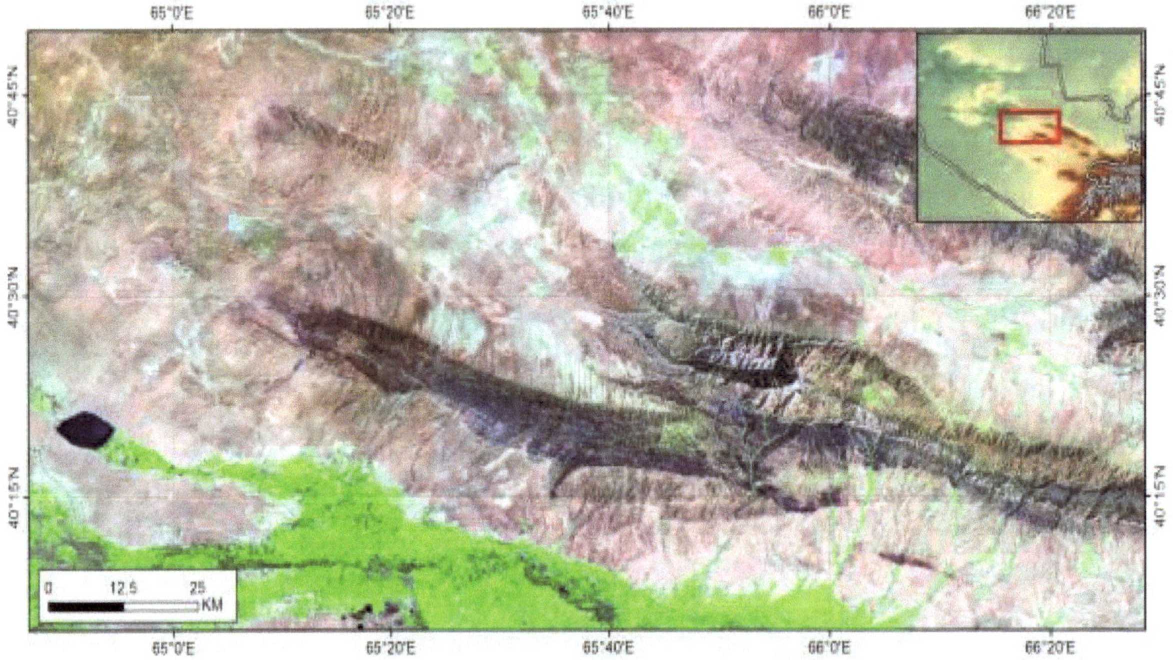

Figure 2.1.1. Landsat space image of the study area

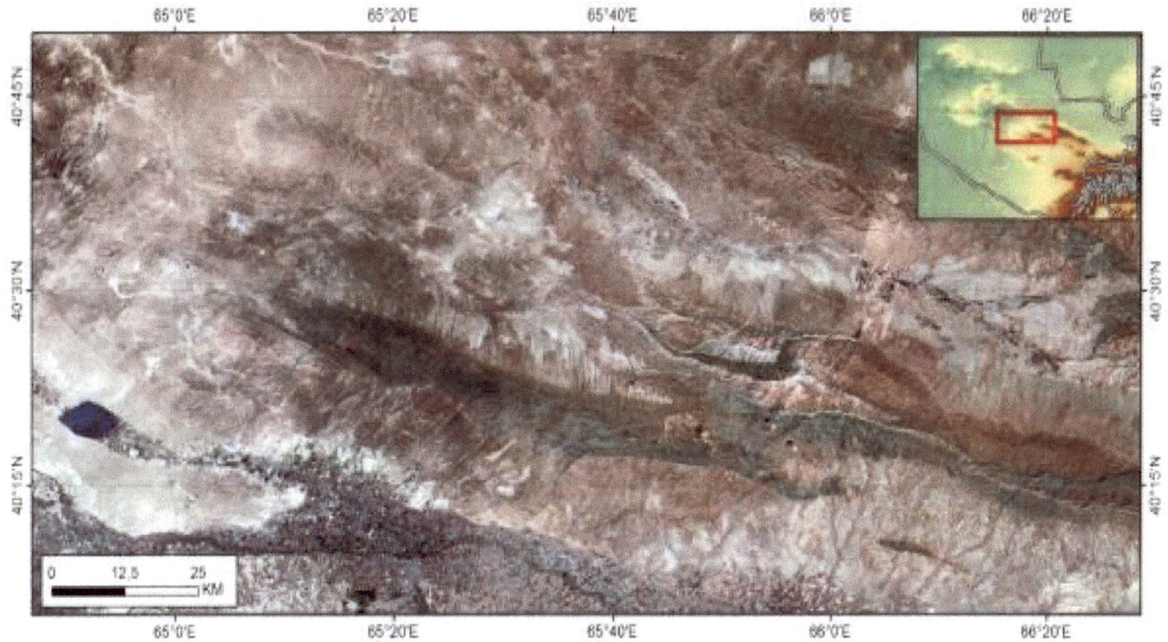

Figure 2.1.2. Aster (TERRA) space image of the research area

In implementation work, relying on special software tools and manuals, it is possible to improve geological information, including in the field of geology, digital space images obtained from satellite systems. The combination of spectral channels of space images made it possible to identify various cosmogeological objects. In the study area, the preliminary data of remote sensing of the Earth was carried out using special software tools (ENVI, ERDAS). The next stage of the work in space photographs prepared for geological decoding and informational enhancement is the strengthening of analytical (decoding) signs of geological objects. For this purpose, the ratio of channels 1, 2, 3, 4, 5, 6, 7, 8 of the Landsat 7 TM images obtained from satellite systems was extensively used. The advantage of the algorithms in the processing of space images is the ratio of spectral channels (Nurkhodjaev et al., 2017, Goipov et al., 2018, 2019).

Using these algorithms, the objects and geological structures that are clearly visible in the top view, which have a positive effect, were separated. A combination of spectral channels was selected in different relativities to create new variants of space images. Principal component analysis (PCA) is often used as a data compression technique. This method of analysis allows placing redundant data into fewer channels. The use of automatic large-scale factual and analytical materials for the process of creating a remote basis requires strengthening the completeness and reliability of the multi-purpose data reflected in the remote basis (MA).

The work was carried out using the available (Landsat 7 TM) 8 spectral channels simultaneously with visual analysis and automatically developed classification of space photo materials. Various processing and use of Landsat 7TM space images increases the geological information in space images by identifying

17

gaps in special structural-decoding complexes. As a result, 5/4, 3/1, 5/7 help to identify other type of objects through relativity algorithm (Glux, 2009). Decoding works (visual and computer) carried out on the basis of preparation of space pictures taken from the satellite system and increase of geological information from them, with the help of deciphering of cracks and annular structures, a comparison of the cosmogeological map of the prospective area in the research area with other data was achieved. Such an approach allows you to focus on the task and solve it correctly. The cosmogeological map uses a remote basis, combined with other maps in the GAT project. In addition, when adding multi-zone channels, the following: space photo materials Landsat 7 TM (channels 1-8) blue (V), green (G), red (R) colors are of great importance for use in the regions of Uzbekistan (Glux, 2009).

In the pre-processing (color and geometric correction) of Landsat 7 TM, 7 spectral channels are used for structure-decoding complexes (SDM), 8 spectral channels are able to obtain more geological information when distinguishing panchromatic, linear and ring structures. Granitoid SDM gives a blue-gray to swamp-colored image in space images. The role of high-resolution space images in increasing geological information is incomparable. Metaterrigenous SDMs are characterized by light blue-green, light brown color in space photographs. Carbonate and siliceous SDM give light blue-gray-brown hues in space photographs. Terrigenous-shale SDM appears in space imagery as a bright blue-green. Carbonate-terrigenous SDM can be characterized by bright gray light gray color. In some places, the changes associated with the marbling of the limestones give the shades from light gray to light brown in space pictures. It was observed that the SDM with olistostrome is expressed in bright gray to dark-swamp colors in space photographs (Togaev, 2018). Yellowish-gray colored gravel, sandstone and siltstone layers in the foothills are represented by red-brown color in SDM. Sandy-clay SDM is represented by bright to light blue phototones, and proluvial deposits are light gray to light blue.

The composite view of R1, G4, B7 channels in the space images of the Southern Nurota area and adjacent areas reflects the natural colors of the areas. Spectral channels R4, G1, B6: 4/3, 5/4, 5/7 (use of different methods), reflecting metamorphic facies of the interrelationship of different channels, if it allows to get more information in the mapping of carbonate, terrigenous, volcanic, volcanic-sedimentary layers, shows Quaternary deposits in channels in bright pink colors. The primary colors of the spectrum are 4/2=R, 7/1=G, 5/6=V and 4/1, respectively, according to their combination and proportion; 4/3; Correlation of 1/3 RGB channels allows effective identification of altered zones, volcanic-sedimentary and mixtite complexes in SDM space images. Landsat 7 TM space imagery was used to summarize the main channels for the study area. Also in adjacent areas: channel 5 (IK) - in red (R), channel 4 (near red and IK) - in green (G) and channel 2 (from yellow to blue) - in blue (B) processed channels are summarized (Shortsman, 2013, Glux, 2009).

High-resolution space images using Earth remote sensing materials, especially on a large scale, distinguish rock formations with their color differences and allow for better mapping. Quaternary sediments formed in the foothills are shown in light gray and yellow colors, and Paleozoic sediments are shown in dark brown and brown colors. Limestones and carbonate rocks scattered throughout the area (Oktov Ridge) are depicted in light gray and yellow-white colors. Thus, in the composite images, sandstones (yellow), vegetation (green) and water bodies (small lakes are dark, deep basins to black) are mapped in clear images.

2.2. Preliminary Processing of Space Imaging Data

In the research area, the preliminary processing of Earth remote sensing data was carried out using specific software tools, including modern computer programs such as ENVI, ERDAS Imagine (Fig. 2.2.1).

During the initial automatic processing of remote sensing data, the following processes are implemented: - image quality improvement; - image synthesis; - image filtering; - cropping pictures in the study area, etc. includes.

Such works were used in almost every processing (Nurkhodjaev et al., 2017, Khachaturyan, Avezov, Glukh, 2011). It can be divided into several types of operations according to processing characteristics. The first type usually includes the value of each individual pixel using a tabular method of representing the regenerative function. It is a descriptive representation of the reproduction of various types of linear and non-linear brightness (contrast), designed to improve the visual perception of information. Image geometric descriptions constitute the next type of operations group. They include cutting out the image or their parts (mosaic), necessary fragments, enlarging the image, bringing the image to cartographic projection (Gonzalez, 2006, Zagubnyy, 2004, Glux, 2009). In the considered processes, it is important to create different color compositions that are optimal for visual perception. This group of reproduction allows obtaining color images in conditional (false) colors, which is one of the methods of processing most data and is important in decoding structural-decoding complexes. This reconstruction is performed using ENVI 4.5 software and imported into ArcGis 9.2 for decoding.

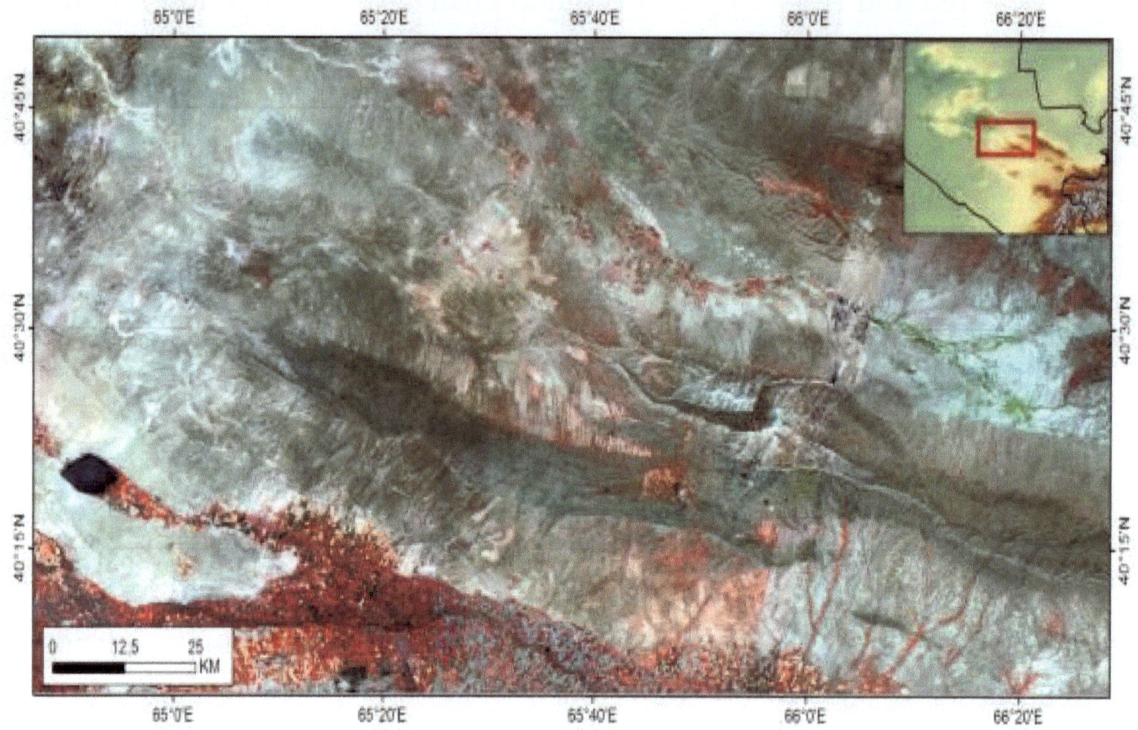

Figure 2.2.1 Generalized view of the study area in Landsat and Aster satellite images

The main characteristics of Earth remote sensing materials are the scale or spatial (latitudinal) resolution of the images, that is, the detail of the image and their types (black-and-white, color, multi-zone, electromagnetic vibrations, infrared or visible spectra), the season and time of mapping are important.

The choice of detail of the photographs depends on the objectives of the research and the geographical characteristics of the object of observation. The first is the size of the object corresponding to the scale of the decryption work; The second is the clarity of the boundaries of the object being decoded, the clearer the boundaries of the objects, the more detailed the image looks.

Since the main task of the research work is to create a cosmogeological map-distance basis and to separate the prospective areas through the data of the conducted geological-geophysical and geochemical studies and the results of cosmogeological data, it was appropriate to use Landsat 7, Aster, Quick Bird -rgb space images (Togaev, 2018 , 2020). Quick Bird space images have the best resolution of all satellites available in the field of Earth remote sensing (except GeoEye and World View-3). The images are both panchromatic (0.61m per pixel) and multispectral (2.4m per pixel) and have a synthesized appearance, with good image decoding properties. Preparation of high-resolution space images for initial decoding involves several stages of preliminary processing (Fig. 2.2.2).

20

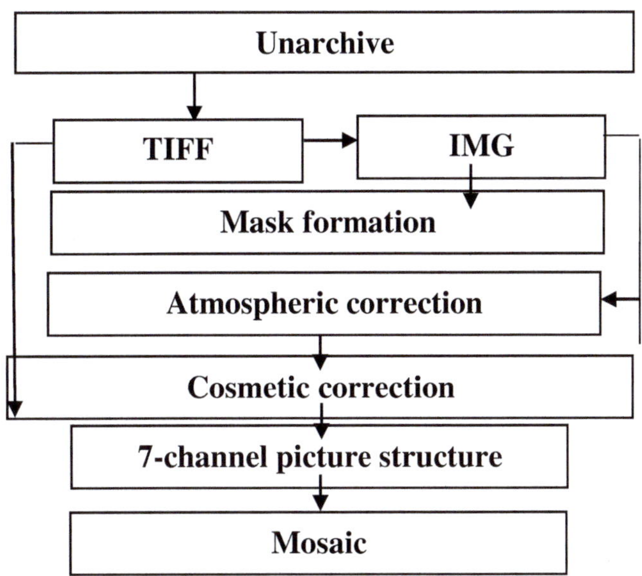

Figure 2.2.2. A drawing of the preliminary processing of space imaging materials

Clouds are hidden using one to six spectral channels, that is, the first, blue channel of the electromagnetic spectrum is depicted in the visible range (0.45 - 0.52 μm (micrometers) and the sixth channel in the thermal range (10.40 - 12.5 μm). It takes a bimodal or multimodal appearance. strongly reflected in the visible range and appears white or gray in the image. At first, the average values for the distribution of luminances in the visible and thermal channels are determined. Luminance values from both channels are then saved for further analysis. Atmospheric editing is performed using the standard method of atmospheric editing provided by Landsat (Pat S. Chavez Jr., 1988, 1989, 1996) (Alkan et al., 2013). Cosmetic correction was carried out to eliminate the visible errors of the sensor work and eliminate the effect of noise (Fig. 2.2.3).

Figure 2.2.3. "Falling" of lines (participation of the row with a zero value of brightness).

Many of these errors can be corrected by applying filters.

7-channel satellite image formation is performed using the "Layerselection and Stacking" module of the ERDAS IMAGINE software. In the module, there are input and output files, spectral channels, called layers, are alternately added and the raster type is specified, the zero value in the image is removed when collecting the space image. The last step is mosaicing the data, this type of work is labor-intensive, requiring simultaneous color editing of seven channels and visual skills.

During the work, additional filtering is carried out, the average values of each pixel are calculated for each channel, the standard deviation of each pixel of the spectral channel is calculated, the photo season is taken into account, and the interferences remaining in the space image are filtered (Kashkin, Sukhinin, 2001). In the process of creating a mosaic from photos, it was often necessary to face the problem that the color descriptions of the photos are not of the same type at a significant level. This is due to the fact that the images used for the mosaic may have been taken by different satellite systems and at different times. Sometimes a photo taken at a close time interval on the same sensor can differ significantly. To solve this problem, the ERDAS Imagine program provides a special function for leveling the color characteristics of the image. This is color balancing or Color Balancing. This function is achieved in the process of creating a mosaic when working in the input image mode. These reconstructions are performed using the ENVI software and imported into ArcGis for further decoding (Kashkin, Sukhinin, 2001). At present, ENVI, ERDAS, ArcGis software products are widely used in "EMZ and GAT-Technology Center", which fully perform complex complex processing of space photo materials. All such regenerations were carried out in full in the area. All surveyed areas were covered by Landsat ETM in 1 space image and 6 Aster space images. In this research work, Earth remote sensing materials, i.e., digital space images taken from satellite systems, were used. Quick Bird imagery has also been widely used to illustrate research. The resolution of Landsat 7 TM space images is up to 15 meters. This system provides representation of cosmogeological objects by describing them in 7 spectral channels, including two thermal channels, and in various combinations (Fig. 2.2.4.-2.2.5).

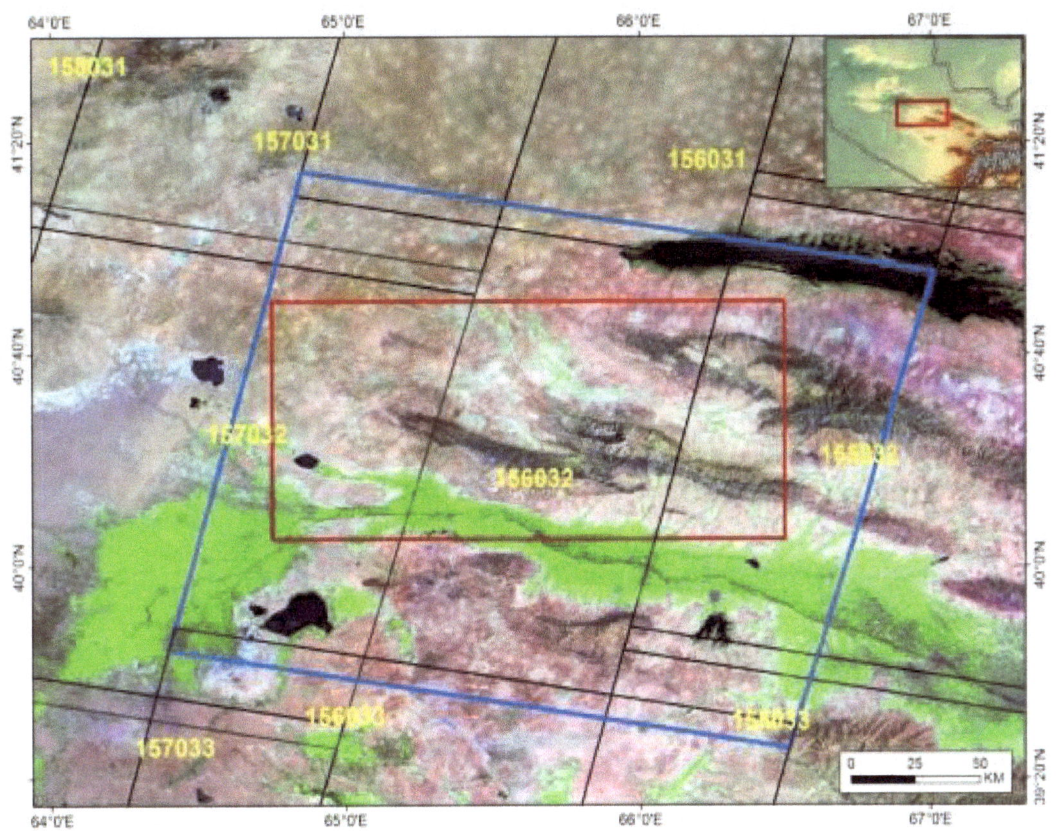

Figure 2.2.4. Location scheme of Landsat 7 TM (1 scene) space image in the study area

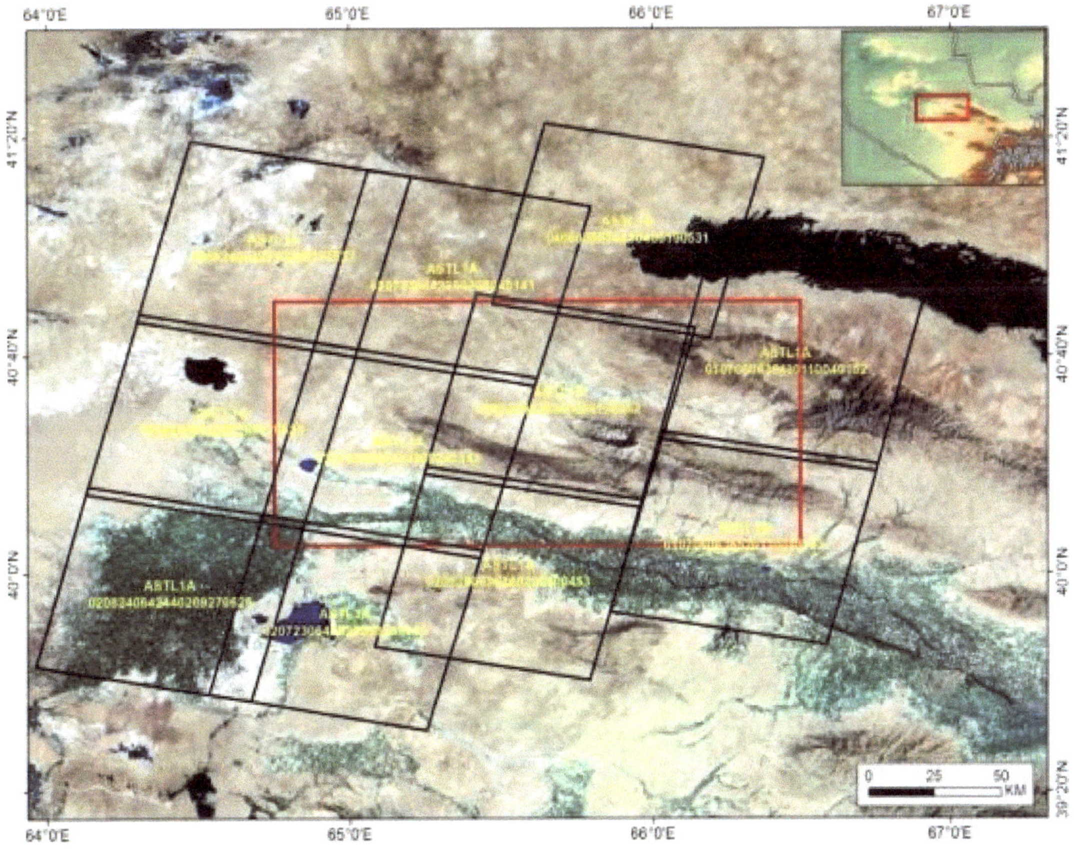

Figure 2.2.5. Location scheme of Aster (scene 6) space photo in the study area

<h2 style="text-align:center">2.3. Geological Decoding of Remote Sensing Materials of the Earth</h2>

Decoding materials of remote sensing of the earth (photographs) means reading, qualitative and quantitative characteristics of geological objects (shape, geometric size, structure, optical properties, laws of interdependence) in the place shown in the photo. An integral part of the use of remote sensing methods in geological research is geological decoding.

During geological deciphering, geological information is obtained not only about the manifestation of geological objects on the surface of the earth, but also about closed structures under plant-soil and sediment deposits (Asadov et al., 2015, Nurkhodjaev et al., 2017, 2019).

Two basic visual and automatic methods are widely used in decoding. In visual decoding, the use of technical tools that expand the capabilities of the eye in time is accepted. Working with multi-zone images requires different methodological techniques for visual decoding of a multi-zone image than a single image. A universal method is to synthesize an image with the choice of a color synthesis variant to optimally solve the specific task of decoding. During decoding, various objects are optimally depicted in pictures in different spectral zones (Nurkhodjaev et al., 2017, 2019). Below are examples of the experiences and samples obtained in the Southern Nurota region, the options for the proportions of these channels and the geological data obtained from them are considered as one of the results of cosmogeological research in the Southern Nurota region. As a result of the conducted work, the digital space images taken from the Landsat 7 satellite were made based on the synthesis and proportionality of spectral channels (Table 2.3.1). In such proportions, the geologist-expert selects the most optimal option in whichever option has the highest level of obtaining geological information.

Table 2.3.1

Synthesis and proportion of spectral channels and geological information in the Southern Nurota area (as an example of Landsat space image)

Landsat 5.7 channel ratio	Geological information	Examples of space photos
	4,3,2. The ratio of these channels is basically "Artificial Color", a standard combination. In red, urban buildings are depicted as green-blue, and the color of the soil is from dark to dark brown. Also, in this image, places with scattered temporary and permanent flowing water deposits, igneous and sedimentary rocks are highly visible.	

7,4,2. This combination gives an image that is close to natural colors, and in the space image, relief forms, their components, and denudation processes are clearly visible. The boundaries of rocks of different composition are separated by differentiated photos. Photo objects of magmatic origin are recorded more reliably. Intrusive rocks are strongly differentiated from sedimentary-volcanic rocks.	
4,5,1. Through these channels, the relationship of igneous rocks to the surface of the earth and the deposits of different periods is well shown in space pictures. In these channels, plants are represented in red, brown, reddish-yellow colors. Adding a red infrared channel in the image allows better separation of rocks of different composition.	
5,4,3. These channels are characterized by providing a lot of information and color brightness to the decoder through their ratio. The composition of rocks, cracks on the surface of the relief, denudation processes, places where alluvial-delluvial deposits are scattered, as well as man-made areas on the surface of the earth are revealed, and closed areas can be used as an indirect criterion for studying the features of the geological structure.	
3,2,1. Through the proportionality of these channels, the geological structures, the appearance of different rocks in the deposits of different periods, discontinuity connections and the appearance of tension are expressed in a high degree in space photographs.	
5,3,1. Through the correlation of these channels, geological structures and textures are clearly distinguished in space images. It can be said that at the same time it allows to separate period beds like channel 7-3-1. In brownish-yellow color, boundaries and relationships of geological structures and intrusive massifs are distinguished and allow obtaining geological information.	

Geological and tectonic structures are evident in these samples. In the decoding of multi-zone images, carbonate deposits reveal structures of flowing color, their brightness is high and their interaction with neighboring structures is well reflected in tectonic disturbances and on the relief surface (Fig. 2.3.1).

Figure 2.3.1. Appearance of Silurian deposits and Oktov intrusive in the southwestern part of the area according to the channel ratio in the Landsat space image.

At the same time, these combinations are important for obtaining additional geological information when identifying separate geological structures in visual decoding (Figures 2.3.2-2.3.3).

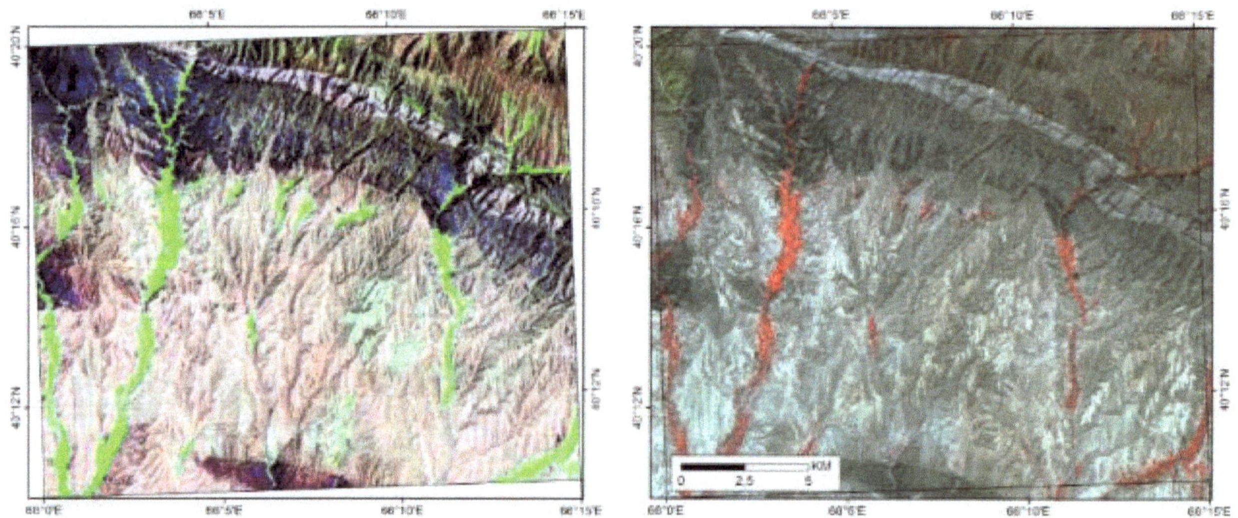

Figure 2.3.2. R7 G4 B2 channel ratio

Figure 2.3.3. Aster space photo
Ratio of channels R1 G2 B3

A table of RGB (red-red, green-green, blue-blue) channel combinations is compiled for the research area. The distribution of these spectral channels in the RGB system serves to ease the user's problems. For example, R3G2B1 and R5G6B4 differ in color descriptions.

The next stage is the allocation of channels by individual decoding symbols, which was formulated as an addition in the research area to solve this task (Table

26

2.3.2). The importance of the ratio of these channels is that in 1:50000, 1:25000 scale works, these channels can be used. Data obtained from aerial photographs is processed based on certain decoding criteria. Decoding criteria: divided into direct and indirect types (Kats, Ryabukhin, Trafimov, 1976, Kats, Poletaev, Rumyantseva, 1986, Kats, Tevelev, Poletaev 1988). The criteria that give an impression of the geological object directly by observing the aerial photographs are called direct criteria.

Table 2.3.2

Landsat 7, RGB space image channel ratio						
№	1	2	3	4	5	6	7
1	R1G2B3	R2G1B3	R3G1B2	R4G1B2	R5G1B2	R6G1B2	R7G1B2
2	R1G2B4	R2G1B4	R3G1B4	R4G1B3	R5G1B3	R6G1B3	R7G1B3
3	R1G2B5	R2G1B5	R3G1B5	R4G1B5	R5G1B4	R6G1B4	R7G1B4
4	R1G2B6	R2G1B6	R3G1B6	R4G1B6	R5G1B6	R6G1B5	R7G1B5
5	R1G2B7	R2G1B7	R3G1B7	R4G1B7	R5G1B7	R6G1B7	R7G1B6
6	R1G3B2	R2G3B1	R3G2B1	R4G2B1	R5G2B1	R6G2B1	R7G2B1
7	R1G3B4	R2G3B4	R3G2B4	R4G2B3	R5G2B3	R6G2B3	R7G2B3
8	R1G3B5	R2G3B5	R3G2B5	R4G2B5	R5G2B4	R6G2B4	R7G2B4
9	R1G3B6	R2G3B6	R3G2B6	R4G2B6	R5G2B6	R6G2B5	R7G2B5
10	R1G3B7	R2G3B7	R3G2B7	R4G2B7	R5G2B7	R6G2B7	R7G2B6
11	R1G4B2	R2G4B1	R3G4B1	R4G3B1	R5G3B1	R6G3B1	R7G3B1
12	R1G4B3	R2G4B3	R3G4B2	R4G3B2	R5G3B2	R6G3B2	R7G3B2
13	R1G4B5	R2G4B5	R3G4B5	R4G3B5	R5G3B4	R6G3B4	R7G3B4
14	R1G4B6	R2G4B6	R3G4B6	R4G3B6	R5G3B6	R6G3B5	R7G3B5
15	R1G4B7	R2G4B7	R3G4B7	R4G3B7	R5G3B7	R6G3B7	R7G3B6
16	R1G5B2	R2G5B1	R3G5B1	R4G5B1	R5G4B1	R6G4B1	R7G4B1
17	R1G5B3	R2G5B3	R3G5B2	R4G5B2	R5G4B2	R6G4B2	R7G4B2
18	R1G5B4	R2G5B4	R3G5B4	R4G5B3	R5G4B3	R6G4B3	R7G4B3
19	R1G5B6	R2G5B6	R3G5B6	R4G5B6	R5G4B6	R6G4B5	R7G4B5
20	R1G5B7	R2G5B7	R3G5B7	R4G5B7	R5G4B7	R6G4B7	R7G4B6
21	R1G6B2	R2G6B1	R3G6B1	R4G6B1	R5G6B1	R6G5B1	R7G5B1
22	R1G6B3	R2G6B3	R3G6B2	R4G6B2	R5G6B2	R6G5B2	R7G5B2
23	R1G6B4	R2G6B4	R3G6B4	R4G6B3	R5G6B3	R6G5B3	R7G5B3
24	R1G6B5	R2G6B5	R3G6B5	R4G6B5	R5G6B4	R6G5B4	R7G5B4
25	R1G6B7	R2G6B7	R3G6B7	R4G6B7	R5G6B7	R6G5B7	R7G5B6

Decoding of aerial images using direct criteria is important in mountainous regions, areas where geological formations protrude from the earth's surface, rocks of different lithological composition, and places where sharp tectonic movement is observed. Indirect signs of deciphering the signs of geological formations by means of natural sources shown in aerial photographs are called. These deciphering signs are the main indicators in the study of the geological structures of low plains, platform lands, intermountain valleys. As a result of the research conducted by E.Barret, A.Kurtislar, when decoding objects, the information obtained from images is based on the following factors: Shape, size, photo, shade, shape, texture, location, expressiveness, stereo effect. Decoding comes from the geological purpose and tasks of remote sensing of the Earth (Asadov, 2011, Barrett, Kurtis, 1979, Nurkhodjaev et al., 2019). In the course of geological decoding, the determination of geological features is based on direct and indirect signs (Fig. 2.3.4). The main unit of the channels was selected based on their brightness and structural characteristics, visually comparing the available proportions of the spectral channels in space images (Table 3.4.3).

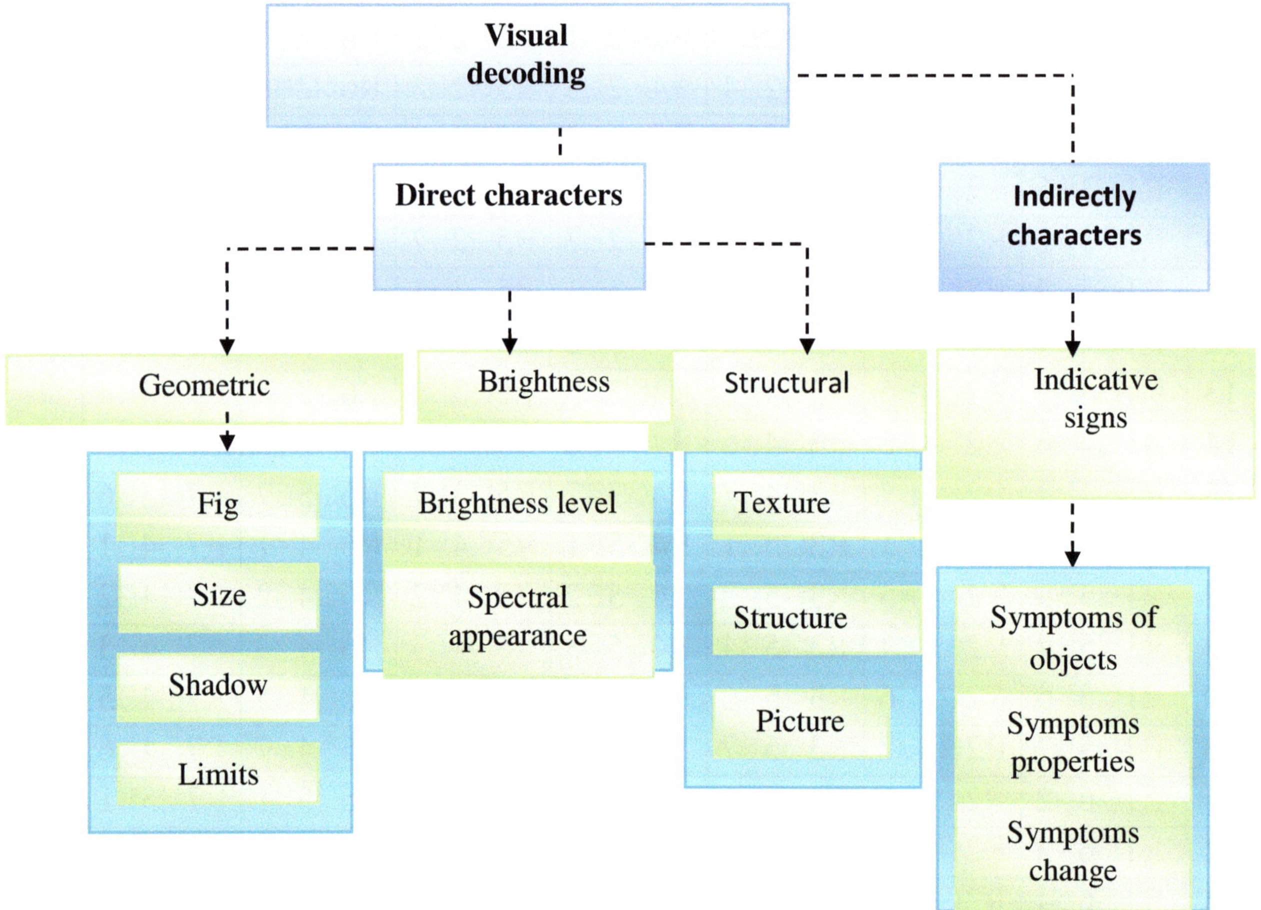

Figure 2.3.4. Scheme of direct and indirect criteria for visual decoding

Selection of channel ratio by direct decoding symbols

Allocation state by decoding symbols		
Geological structures	**For vegetation and greenness index**	**In hydrogeological studies**
R 7 G 3 B 5	R 5 G 3 B 1	R 6 G 4 B 7
R 4 G 3 B 5	R 4 G 3 B 5	R 1 G 3 B 7
R 6 G 2 B 3	R 1 G 3 B 7	R 1 G 5 B 2
R 3 G 1 B 7	R 1 G 4 B 6	R 1 G 4 B 3
R 1 G 4 B 7	R 1 G 3 B 6	R 1 G 5 B 3
R 1 G 4 B 6	R 1 G 4 B 5	R 1 G 5 B 5
R 1 G 3 B 4	R 1 G 3 B 4	R 2 G 3 B 7
R 1 G 3 B 6	R 1 G 5 B 3	R 3 G 2 B 5
R 1 G 3 B 5	R 1 G 5 B 7	R 4 G 1 B 5
R 1 G 5 B 4	R 1 G 5 B 6	R 4 G 3 B 1
R 1 G 5 B 6	R 3 G 2 B 5	R 4 G 3 B 5
R 4 G 3 B 5	R 3 G 2 B 5	R 4 G 5 B 7
R 4 G 3 B 6	R 4 G 2 B 7	R 4 G 6 B 3
R 4 G 5 B 2	R 4 G 3 B 5	R 5 G 1 B 2
R 4 G 5 B 3	R 4 G 5 B 7	R 5 G 1 B 7
R 4 G 6 B 3	R 4 G 6 B 7	R 5 G 2 B 3
R 4 G 6 B 7	R 5 G 3 B 7	R 5 G 2 B 7
R 5 G 2 B 1	R 5 G 4 B 1	R 5 G 3 B 3
R 5 G 2 B 3	R 5 G 2 B 7	R 5 G 3 B 7
R 5 G 2 B 7	R 5 G 3 B 5	R 5 G 4 B 2
R 5 G 1 B 6	R 5 G 4 B 2	R 5 G 6 B 1
R 5 G 6 B 3	R 5 G 4 B 7	R 5 G 7 B 6
R 5 G 6 B 7	R 6 G 3 B 4	R 6 G 1 B 7
R 6 G 2 B 3	R 6 G 4 B 1	R 6 G 2 B 7
R 6 G 1 B 3	R 6 G 5 B 1	R 6 G 3 B 5
R 6 G 2 B 1	R 7 G 2 B 5	R 7 G 2 B 6
R 6 G 2 B 4	R 7 G 3 B 1	
R 6 G 3 B 1	R 7 G 2 B 5	
R 6 G 3 B 7	R 7 G 2 B 3	
R 6 G 4 B 3	R 7 G 4 B 2	
R 6 G 3 B 5		
R 6 G 5 B 2		
R 6 G 5 B 7		

R 6 G4 B 2		

The formation of different proportions of channels is used in practical visual decoding. The advantage is aimed at solving practical tasks conveniently and quickly depending on the user's direction (Togaev, 2018). Automatic methods also occupy a special place in solving practical tasks during automatic decoding (Nurkhodjaev, 2017, Pirnazarov, Asadov et al., 2017).

SS (color composition), PSA, Mincomp, Hydrocomp, Kirsh, Sobel and other methods of space image processing were used in the research work (Nurkhodjaev et al., 2017, Ergashev, Isakhodjaev, Asadov, 2007). For the automatic detection of linear structures, space images can be identified by infrared spectral channels, and software such as Winlessa and Alina using complex algorithms have been created (Goipov, 2019., Zlatopolsky et al., 2004, Krutikov, 2015). Linear objects in space photographs are considered lineaments, and the term was introduced by the American researcher V. Hobbs at the beginning of the century (1911). They are manifested by cracks on the earth's surface or indirectly by geological and landscape anomalies. In visual and automatic decoding, delimiting lineaments appear in pictures with a much higher expression (Asadov et al., 2015, Borisov, Glukh, 1982).

The importance is the fault systems separating the main geostructures. Automatic decoding is carried out on the basis of special modern computer programs and is important because it provides the researcher with new information that was not detected or ignored by the observation method. Some of its shortcomings are corrected by the geologist during the work process. Therefore, on the basis of computer technologies, the process of decoding materials of remote sensing of the earth should be carried out under the strict supervision of a geologist-expert (Avezov, 2004, Nurkhodjaev, 2017). The results of this method are presented below in Table 3.4.4. These obtained results are considered as the main parameters in the automatic representation of linear geological structures built on the basis of the above-mentioned automatic methods of space images, an elongated object or a structural-decoding complex (Togaev, 2018).

Geological informativeness of decoding multispectral space images by automatic decoding in infrared channels

Meth ods	Samples for special spectral processing to obtain geological information		
Kirsh			
Sobel			
PCA			
ITS			
MinComp			
HydroComp			

The Erdas Imagine program, which is widely used in solving similar geological tasks in the world experience, was used for the manifestation of linear structures in Landsat 7 images and their processing. As a result, lines corresponding to geological structures were obtained (Figures 2.3.5-2.3.7).

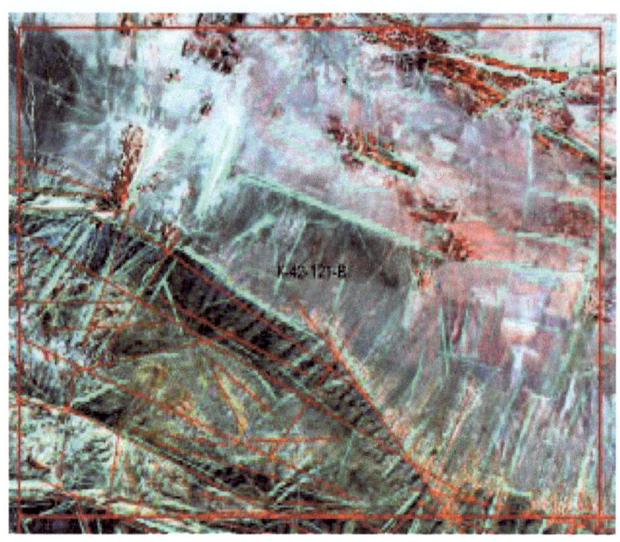

Figure 2.3.5. Manifestation of linear structures depending on the shape of the relief in the southwestern part of the Southern Nurota region

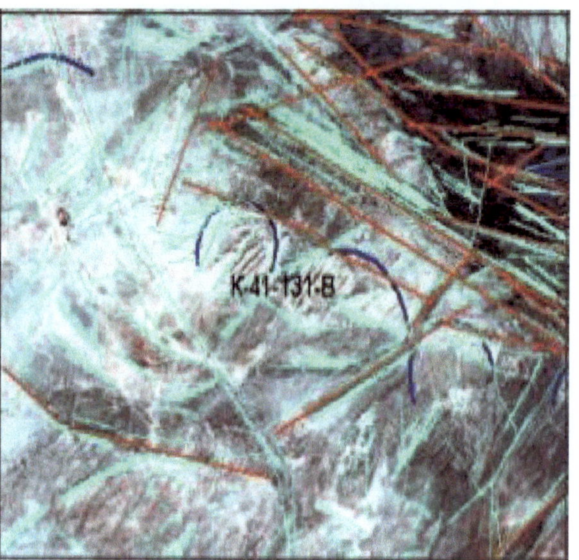

Figure 2.3.6. In the eastern part of the region, separation of linear and ring-shaped structures for sections with sharply changing terrain

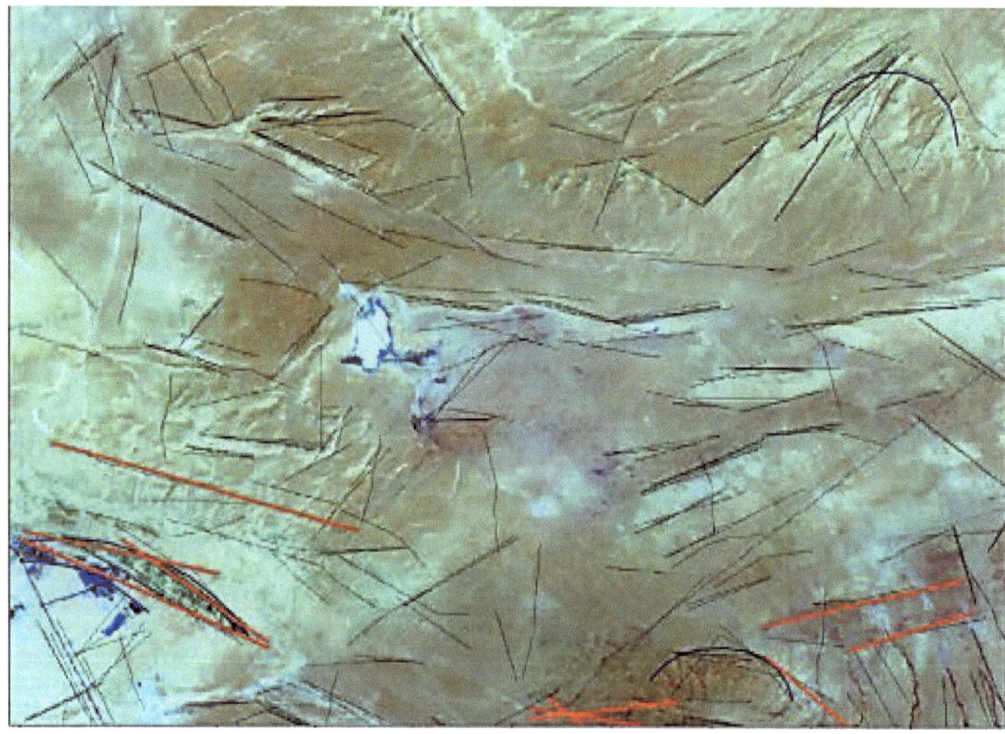

Figure 2.3.7. Separation of ring-shaped and linear structures in the northeastern part of the region based on the processing of the digital relief model

Linear structures are usually evident as striated anomalies, with partially discernible and convergent boundaries on platform-like planes, and represent lithosphere heterogeneity. If the space image is high and low in latitude, they can image deep structures using bright spectral reflectance. A comparison of Aster's nighttime thermal images with Landsat images is used for high resolution. The tasks of finding closed geological structures can be solved due to the change and comparison of thermal characteristics of the earth's surface in the above images (Yakimov, Krutikov, 2015). Red to dark red indicates high surface temperatures, yellow indicates moderate, and green and orange indicate low temperatures. The boundaries of structures are divided according to thermal properties. Cracks are distinct at much lower temperatures, clustered between higher temperature structures, some of which appear as linear structures. Ring-shaped structures of different scales are displayed along the spectrum limits of yellow colors. Determining the informativeness of geological structures is one of the most difficult tasks of automatic decoding (Kats et al., 1988). In the visual decoding of pictures, it is necessary to identify and separate objects based on decoding symbols. It is desirable to make comprehensive use of the achievements in the field of visual and automatic geological decoding in obtaining information on space photography. Let's look at the results of the above methods. The combination of channels makes it possible to get an image of the place from blue to brownish-red, geological-landscape objects, which record the distribution of separate formations of rocks between them (Fig. 2.3.10). Principal component analysis is a technique for analyzing multispectral data comparisons. Data comparison - here means that when the value of pixel brightness increases in one spectral channel, it also increases in other spectral channels.

Figure 2.3.10. RGB ➜ Separation of field SDM and major linear tectonic elements based on a standard combination of 7,4,2 channels

In the PSA method – in this method of space photo processing, it shows the main constituents of the geological structure in the field. Rocks of different ages and different types are photographed differently in the space image and look good (Fig. 2.3.11).

Spectral signs in the two-dimensional space, with the increase in the value of the first channel, it also increases in the second, which means a high correlation between these channels. Such images are in different colors and allow to map rocks of different ages (Ergashev et al. 2007).

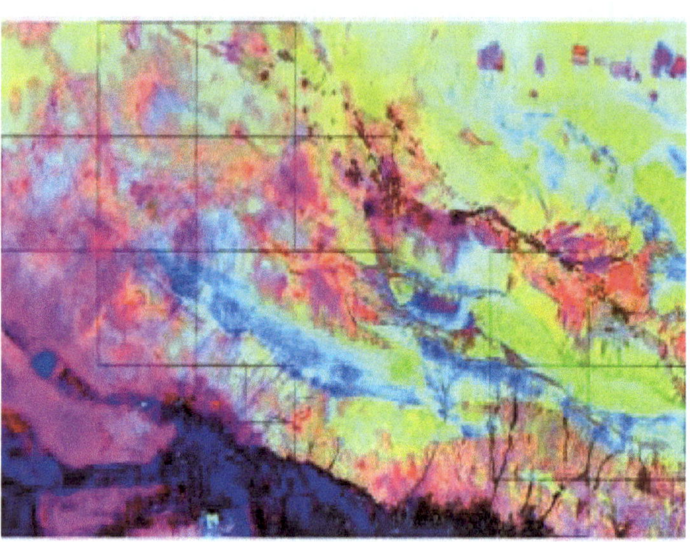

Figure 2.3.11. Space image processed by PSA method

Method INS (intensity, hue and saturation). The images taken in the mapping of rocks of different ages and different types, independent of the landscape-climate, are of great importance. It is known that the human eye is not capable of distinguishing many colors, but it is able to evaluate the priority color perception (hue concept) and pure color (saturation concept). In general, color is determined by 3 parameters. **Determines** the degree of intensity and darkening of colors and corresponds to the amount of reflected energy. Hue Determines the dominant natural color of the surface reflecting from the geo-object. **Saturation.** Pure color is scaled from center to edge on the default color bar and is scaled from 0 (gray) to 100% (pure, refined). The results of geological data obtained by these methods are reflected in the works of our country and foreign researchers (Y.P.Pitev, B.Busygin S.Robila). The results of these methods are called indices or new channels. All indices have absorptive and reflective properties. From a geological point of view, these indices determine the difference between different rocks.

The **MinComp method** (mineral constituents method) is based on color composites of 3 indices, namely: clay rock index, iron-bearing minerals and iron oxide index (Kronberg, 1998). As a result, the red color in the image corresponds to clay rocks, the green color corresponds to

34

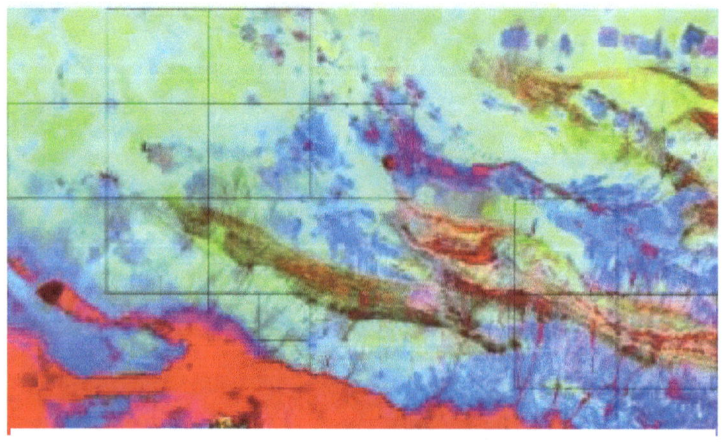

Figure 2.3.12. Space photo processed by MinComp

iron-containing minerals, and the blue color corresponds to iron oxide rocks (Figure 2.3.12).

Processing of space images in the **HydroComp method** is similar to the previous one, based on composites of three indices, only iron oxide and relative index are used instead of iron-containing minerals and iron oxide. In the final processed image, red color corresponds to clay rocks, green to iron oxide rocks, and blue shows differences in albedo effects (Figure 2.3.13).

The **Kirsch method** allows to study the structures closed with relatively small thicknesses (Fig. 2.3.14).

Linear, concentric, arcuate, and isometric structures within folded mountain units, mountain depressions, and subalpine plains are best seen in images of reference areas.

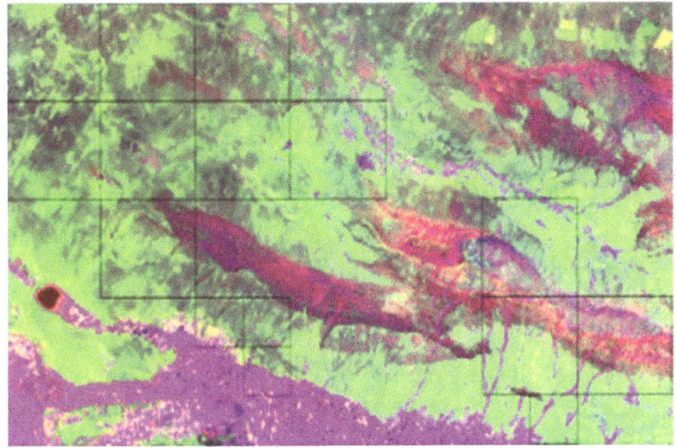

Figure 2.3.13. Space image processed by HydroComp method

Figure 2.3.14. Space photo processed by Kirsch method

Their border is observed in the form of black-and-white, sometimes united lines of different lengths. The densest network of lineaments, the opening of the Paleozoic foundation, is determined in open and semi-closed conditions of the surface. As you move away from the mountains, the resolution decreases and the number of structural units displayed decreases dramatically.

The **Sobel method** of space image processing makes it possible to record eruptive objects far from folded rock formations.

The "Sobel" method is an improved version of the "Kirsch" method, which clearly shows structural diversity, especially in closed areas. Using this filter, it is possible to map the maximum number of structural objects independent of landscape-climatic and geological conditions in the researched areas. Surface and closed cracks are represented by white and black lines as double.

Figure 2.3.16. CC-processed space image of the Southern Nurota area

Figure 2.3.15. A space image of the area processed by the Sobel method

The image of a deep fault is very _______ solid line as previously drawn, but a short cuneiform zone with lines that do not intersect each other in width and length. It helped to map the maximum number of structural objects independent of geological conditions. It has the appearance of concentric structures, lineaments and tectonic cracks (Geological Dictionary, Petrov, 2012). Small, multi-zonal concentric and various structures appear in the zone of large (deep) earth cracks (Fig. 2.3.15).

CC method (color composition) "artificial colors" - it presents the image of the earth's surface in a certain color. In a space image, it is obtained by summing certain spectral channels (Fig. 2.3.16). In the decoding of a space photo, similarity is compared to the image of a known geological object. Visualization of decoding information is carried out in the methods of reproduction of pictures, as mentioned above (Kronberg, 1998, Ergashev et al., 2009). Although the results of automatic processing of space imaging materials do not fully solve the tasks set before geologists, they provide some relief. The main task of processing space image materials is to sort all the separated structures in a sequence, as well as to separate the displayed structures by their spectral characteristics by sorting them.

Summary. The use of Landsat, Aster, Quick Bird space images has created a high possibility for geological decoding of remote sensing materials. As a result, the use of high-resolution space images became important in solving problems of practical geology and in studying the relationship of geological structures with each other; In solving geological problems with the help of Earth remote sensing materials, their most effective RGB channel system (Red-red, Green-green, Blue-blue) ratios were determined. As a result, the use of digital space images made it a reliable opportunity to obtain geological information through the combination of spectral channels; It was found that solving geological problems in the materials of remote sensing of the Earth in the Southern Nurota area, their visual and automatic decoding methods give good results. As a result, the results of complex decoding in the research area gained practical significance in regional geological research; In studying the material composition and tectonic characteristics of geological structures in the Southern Nurota area, the convenience of automatic decoding methods was found. As a result, it was confirmed that the possibility of obtaining geological results from space images with high potential in geological decoding works is effective.

CHAPTER III. REMOTE BASIS AND STRUCTURAL-DECODING COMPLEXES OF THE SOUTH NURATA REGION

3.1. Creation of Remote Basis and Description of Structural-Decoding Complexes

The study of structures and geological bodies with different composition using the materials of remote sensing of the earth on the basis of structural-decoding complexes is of great importance in geological prospecting. Modern digital space photo materials obtained from various satellite systems of the earth's surface were widely used in the conducted research. Having the materials of remote sensing of the earth and decoding it on the basis of special computer programs and direct observation led to the emergence of a new product of cartography of geological research - remote basis (MA) (Togaev, Nurkhodjaev, 2019, 2020).

Remote basis - belongs to the new generation of cartography, it is a new independently created map, combining new data obtained from space images in regional geological studies with the results of previously conducted research. Its advantage is that it summarizes all the geological, geophysical and geochemical data involved in the geological structure of the research area based on a completely different approach. It is worth noting that the remote basis, which is part of the new generation maps of the cartographic process, is a product of a new direction in the field of geology - aerocosmogeological methods or remote geology.

Also, this map represents a new level of organization of data of previously unknown or unused factors from non-traditional methods of studies involved in mapping processes of large areas, while acting as an independent document in regional geological studies. When it became possible to take digital photos of the Earth, a new era of using space photo materials (CSM) - a cosmogeological map - the era of creating a remote basis began. Remote base interpretation allows to analyze the materials collected during regional geological exploration and summarize them in space images (Nurkhodjaev et al., 2017, 2019, Snitka et al., 2016, Togaev, 2018). In this research, for the first time, the results of the remote base cosmogeological research in the Southern Nurota region, the results of geological-geophysical and geochemical data were summarized.

The materials are combined using modern GAT-technologies, allowing to summarize data on cracks, structural-decoding complexes of different genesis and content, ring-shaped structures, cosmogeological objects of point, line, area descriptions and information obtained by traditional methods. These geographic information systems (GIS) are a system that provides collection, storage, transformation and management of spatial geographic data using modern technical tools and special tools.

After collecting a set of satellite images (Landsat, Aster, Quik Bird, SRTM, etc.) for the study area, they were re-linked to the geographic coordinate system (Lat/Lon, WGS). Additional thematic channels were created using Erdas Imagine software. As a result, data-rich space images were formed for visual and automatic decoding. This was done using automatic decoding classification of space pictures. Classification was performed in the analysis of the study area and 16 signature classifications were used. Structure-decoding complexes (SDM) were separated according to the results of analyzes and space photographs. SDM, tectonic structures and annular structures were distinguished. Their location in space (latitudes) and mutual relations were demonstrated (Goipov, 2019, Nurkhodjaev, Togaev, Ibragimov, 2011). It involves comparison, comparison and their logical interpretation (visual and automatic) in the creation of a distance basis. The interpretation of the remote basis is the result of deciphering works based on cosmogeological and field materials (Togaev, 2020).

Currently, few of the projects in this direction are dedicated to field work. Nevertheless, due to the lack of regulation in the methodological-constructive document and the volume of field work in thematic projects is decreasing (Ergashev, Asadov, 2001). In recent times, as a result of the increase in the possibilities of technical tools and photos for remote sensing of the Earth, as a result of the improvement of the methods of geological decoding of these materials, a new technological drawing of the decoding of the materials of remote sensing of the Earth was developed as a method and recommended for practice (Certificate t/r 001951).

It should be emphasized that the main idea is based on the principle of simplicity from generality, using different scale series of different space pictures at the same time. Available information based on the results, taking into account geological-geophysical, geochemical and other factographic data, a preliminary deciphering interpretation scheme with distinct linear and concentric elements of the landscape was created and clarified in visual observations and field studies. The final processing of the received data was carried out mainly in the camera period. At the same time, in the additional study of the pictures of different scales, the results of decoding were taken into account, associating geological observations with geophysical and geochemical anomalies, maps, schemes, decoding symbols and other criteria. Many natural and man-made factors are taken into account when interpreting pictures of the Earth's surface and obtaining geological data, namely, the transparency of the atmosphere, the type of relief, the thickness of the vegetation cover, the presence of settlements, etc. The density (thickness) of the secondary vegetation covers observed in the foothills of the Southern Nurota area caused significant difficulties and has a man-made appearance (Figures 3.1.1-3.1.2).

Figure 3.1.1. Vegetation cover, anthropogenic influence and transition boundary to geological openings in space photographs

Figure 3.1.2. A remote base map of the northeastern part of the region and its comparison with space photography

As can be seen in this picture, it may not coincide with the boundaries of geological objects. Deciphering such areas is difficult due to the fact that a large amount of research covers all areas. This task was solved by choosing a combination of spectral channels of space images in the collective interpretation, obtaining geological data in the installation of signs (indicators) of objects with the involvement of geological and geophysical data. When using this method of decoding, it should be remembered that the same or similar, especially ancient

40

geological formations can appear differently in the landscape. Phototus is the main decoding symbol of 1:50000 scale space photographs (Trofimov, 1986, Nurkhodjaev et al., 2017, 2019). For example, intrusive rocks are deciphered with a flat photoshoot, and in large-scale photos, sometimes a fracture system is deciphered according to its origin. Forms in the form of weak or soft, flattened domes, low hills or flat surfaces are formed (Fig. 3.1.3).

Figure 3.1.3. Space photo of the Oktov intrusive in the South Nurota area (left) and the constructed remote base (right)

In space photographs, shales occupy an intermediate position between shales and granites in decoding. Massive, one type of gneiss is indistinguishable from intrusive rocks by relief and other features (Nurkhodjaev et al., 2019, 2020). Shales form sharp rock formations like granites in mountainous regions or, on the contrary, soft ring-shaped formations. The figure below shows two types of shales exposure images, distinguished by their spectral characteristics (Figure 3.1.4). Geological objects are characterized by different spectral characteristics. In solving a specific geological task, the correct selection of Earth remote sensing materials is very important.

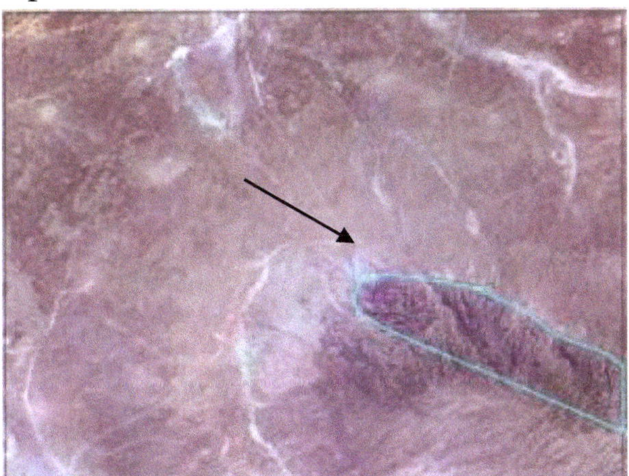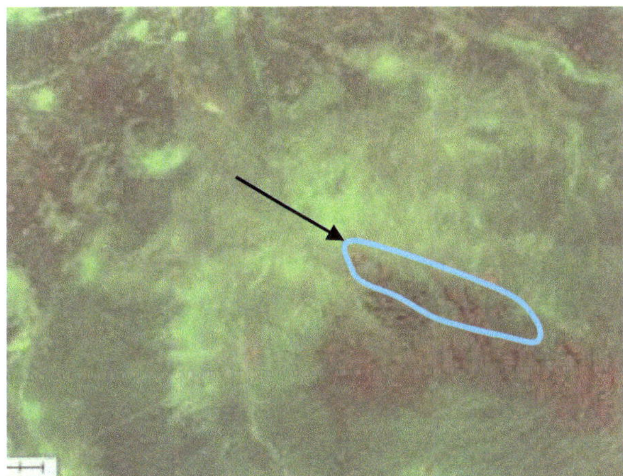

Figure 3.1.4. Appearance of shales in the northern part of the Southern Nurota area in the original (left) and reprocessed space image

The physical properties of sandstones and siltstones depend on the content of cement in them. Their larger counterparts usually have a curved structure and sharp relief forms - sharp-angled peaks, jagged slopes. It is weak in terms of hardness, but rather soft positive (elevations) forms in the relief create flat, wavy or curved surfaces with different steepness of the slopes. In general, conglomerates, sandstones, and siltstones are deciphered reliably enough in space photographs - without being divided into separate types. Limestones and dolomites stand out clearly in space photos. Round and oval forms of karst characteristic of the surface are often formed in areas where carbonate rocks are scattered. Compared to other deposits, alluvial deposits are much easier to manifest. Alluvial deposits are clearly visible in space images and their geomorphological features can also be determined (Sarma, 2011). The geomorphological elements of the river valleys are mainly studied: river beds, banks, terraces, slopes. Riverbeds are reliably deciphered. The alluvial deposits, which are further from the surrounding sections, are sharply distinguished by their grayish or white color (Fig. 3.1.5). The channels of the rivers are stretched along the riverbeds. Old riverbeds are best seen in space photographs, emphasizing the change of the slopes, as well as the linear arrangement of trees and shrubs. In the area, the fields are characterized by vegetation and flowing phototus.

In some cases, it is possible to distinguish between the bottom and the top in the pictures. The terraces extend over considerable distances.

Figure 3.1.6. Demarcation of the boundaries of SDM manifested in the northern part of the territory

Their surfaces are more or less covered with vegetation, shrubs and trees and have an uneven surface. The morphostructures

Figure 3.1.5. South of Oktov mountain separation of alluvial formations manifested on the slope

shown in the relief serve as the main deciphering signs in the identification of field geological objects, geological bodies and the material composition of rocks (Fig. 3.1.6).

Based on the representation of different geologic structures from space images according to different symbols, a preliminary distance basis was created. Various regional geologic data useful for disjunctive faults were extracted for deciphering space images. Despite the century-old history of using this concept in the study of the deep geological structure of the Earth, there is still no universally accepted concept of "lineament". Lineaments are straight or weakly curved natural features of the landscape. They often reflect the fact that the lithosphere is not linear, namely cracks in the earth's crust, flexures in the sedimentary cover, zones of sharp changes in geological structures, high gradient zones in geophysical fields (Asadov et al., 2015, Togaev, 2018, 2020). In geological practice, lineaments can represent channels of transmission of various fluids and solutions, that is, they can serve as direct indicators in the search for mineral deposits. Direct signs were used in the representation of the radar space image, that is, the lines of the object, the size, density of the photo, etc., in addition, lineaments were detected in the stretches of continuous and ring-ring zones with a sharp decrease in the distance between the lineaments (Fig. 3.1.7).

Figure 3.1.7. Linear and ring-shaped structures that were automatically revealed in the eastern part of the region and on the northern slope of Mount Oktov

As indirect signs for closed areas, direct photos are taken from the straightened sections of riverbeds and river valleys, natural transitions of watercourses and ravines-swamps, slopes, vegetation along a straight line or close to it, strong thickening of the vegetation cover and junctions of thick layers of different composition. or used a color shift. The high resolution of the space image reveals large linear structures in them (Fig. 3.1.8).

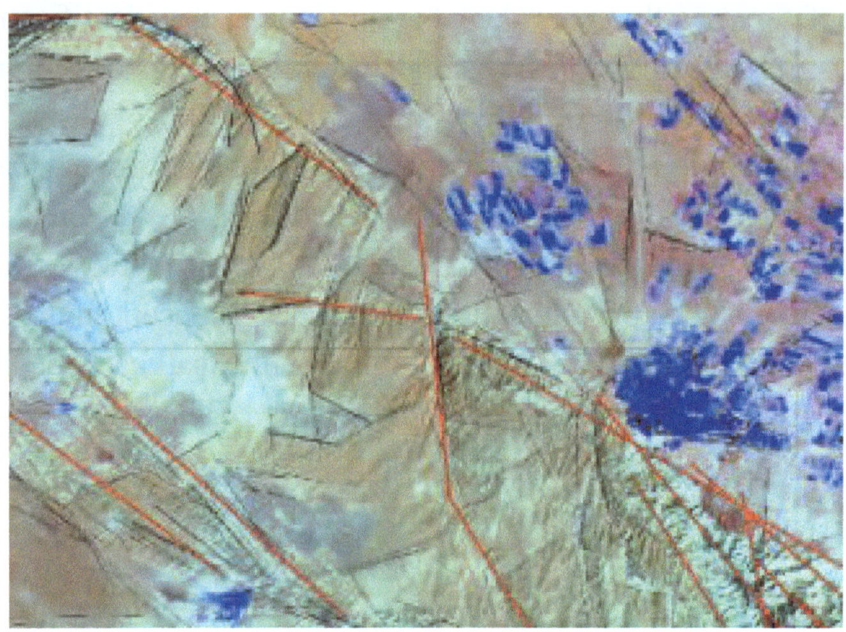

Figure 3.1.8. Separation of large linear structures in the Zarqaynar intrusive massif

After drawing up the initial schemes of linear structures, for their use in further distance bases, the data were re-checked in field conditions and additions were made to the final scheme of linear structures.

All the results obtained from the decoding of the space image and the field work - earth cracks, SDM and their boundaries were shown as a result map based on the distance basis of the Southern Nurota region.

Evaluating the large-scale created remote basis and the role of Earth remote sensing materials, it should be noted that this method in combination with geophysical, geochemical and other methods is of great importance in solving the main tasks facing geologists in performing regional geological work. In this area, it was accepted to introduce the mixtite complex as a new term in cosmogeological research within the SDC. The term mixtite assemblage refers to chaotic assemblages formed as a result of transport of fragmented materials by tectonic, gravitational, and magmatic means. Mixtite complexes are represented by special conditional symbols on a distance basis (the color changes depending on the age of the complex).

Structural-decoding complex (SDC) - geological nature of objects, photoimage, spectral brightness, saturation of the image are separated as different objects with the description of geological structures in the materials of remote sensing of the Earth (A.Q. Nurkhodjaev, I.S. Togaev, 2017).

Magmatic derivatives. The intrusive formations of the Southern Nurota ridge were thoroughly studied by I.M. Isamuhamedov (1955), I.H. Hamroboev (1958), Y.I. Loshkin, Z.A. Yudalevich and other authors. Z.A. Yudalevich and others (1975)

studied the intrusive formations of the area (Stratifitsirovannye i intruzivnye……, 2000, Pyanovskaya, 1984).

Among the intrusive formations, granodiorites forming Chettik, Karatov, Zarkainar, Bitab and Oktov intrusives are leading. Ultrabasic, basic, alkaline-basic rocks are less common and form significant dykes. All magmatic formations were formed during three tectonic-magmatic cycles (Caledonian, Hercynian, Alpine), but the characteristics and manifestations of each are not the same. The Hertzian cycle is considered the most significant in terms of appearance, and the main intrusives - Zarkainar, Bitab, Karatov, Chettik, Oktov - were formed (Pyanovskaya, 1984).

The intrusive rocks in South Nurota and its neighboring areas differ in terms of their composition, bedding conditions, and facies. Irrespective of their composition, intrusive formations have similar structural-deciphering signs, which we classified as granitoid SDM. Oktov and Zarqaynar massifs are characterized by dark-gray colors. Zarqaynar and Oktov granitoid massif and its geological structure are recorded in space photographs - clearly, gray-gray, dark-blue (to white) in the proportion of channels.

Granitoid SDM. Large massifs of granitoids are exposed in the eastern and western parts of the Southern Nurota region. Vein rocks are composed of fine-grained aplite granites and aplite. The Zarqaynar intrusive massif was opened on the western slopes of the Southern Nurota area (Fig. 3.1.9). It is oval in shape, elongated in the sublatitudinal direction and characterized by ring-like structure. In space pictures, this complex is depicted in shades of gray, from liquid to dark green.

Figure 3.1.9. Spatial appearance of the Zarkainar intrusive massif embodied in the Southern Nurota area (right) and decoding results (left)

Mixed complexes. Chaotic (disordered, mixed) complexes (mixtites) have developed in all folded regions of the Earth and are associated with sedimentary and tectonic processes. According to M.G. Leonov's classification (M.G. Leonov

"Olistostromes of structures of folded regions", 1981), the class of mixtites is divided into several types: - gravity-specific collapsing-slide mixtites - olistostromes; - tectonic-gravitational mixtites - tectonized olistostromes; - tectonic mixtites - mélange, tectonic breccias.

Olistostromes are relatives of mixtite formations, the development and origin of the Tien-Shan cover-fold model in Western Uzbekistan, tectonic (mélange), gravitational, collapsing-landslide (olistostrome) and mixed mixtites or chaotic complexes began to be distinguished. SDC with olistostrome, S2-3 developed in the central parts of the South-Nurota Ridge. Shales with lenses of marbles are composed of siltstones. In the Southern Nurota area, this complex is known as Darasoy suite. Layered shales, mica-quartz limestones, dolomites are composed of marbles and marbled limestones. In the space photo, it has a color from dark gray to dark blue in mostly dark colors (Fig. 3.1.10).

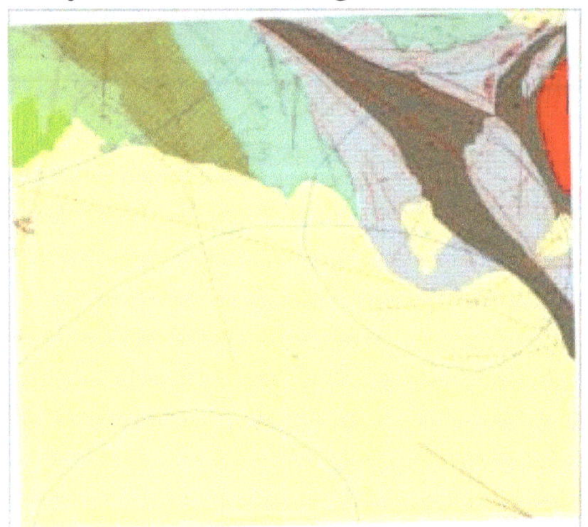
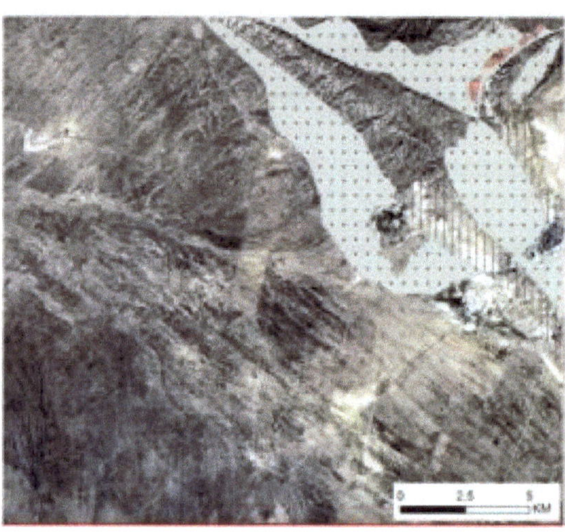

Figure 3.1.10. Spatial representation of mixtite complexes in the Southern Nurota area (right) and decoding result (left)

Silica-carbonate SDC, RR2. The complex is presented in the area as Suliksoy, Sultanboy, Bogambir and Suyaltash suites. The Sultanboy suite was discovered in the form of fragments in the Southern Nurota ridge (Sultanboy, Temirchik wells, Gobduntov mountains, etc.). It is composed of microquartzites and dolomites. It is 100 meters thick. Contacts are tectonic. This siliceous-carbonate SDC is distributed in the southwestern part of the Zarqaynar intrusive massif. This complex consists of siliceous-quartz and dolomites (R.S. Khan, 2006). It is described in gray and dark-gray in the space image. Terrigen-carbonate-shale SDM, €-O. From the Gobduntov Mountains in the east to the Kokcha Mountains in the west, it is spread over almost all hill sections in the region and extends in the form of a narrow tectonic fold (Fig. 3.1.11). Their contacts with the indicated formations are tectonic. Terrigenous-carbonate-shale SDC is composed of terrigenous shales with lens-like layers of siliceous rocks, marbled limestones, argillites, sandstones, and sometimes algal limestone   dolomites   (Togaev,   2018).

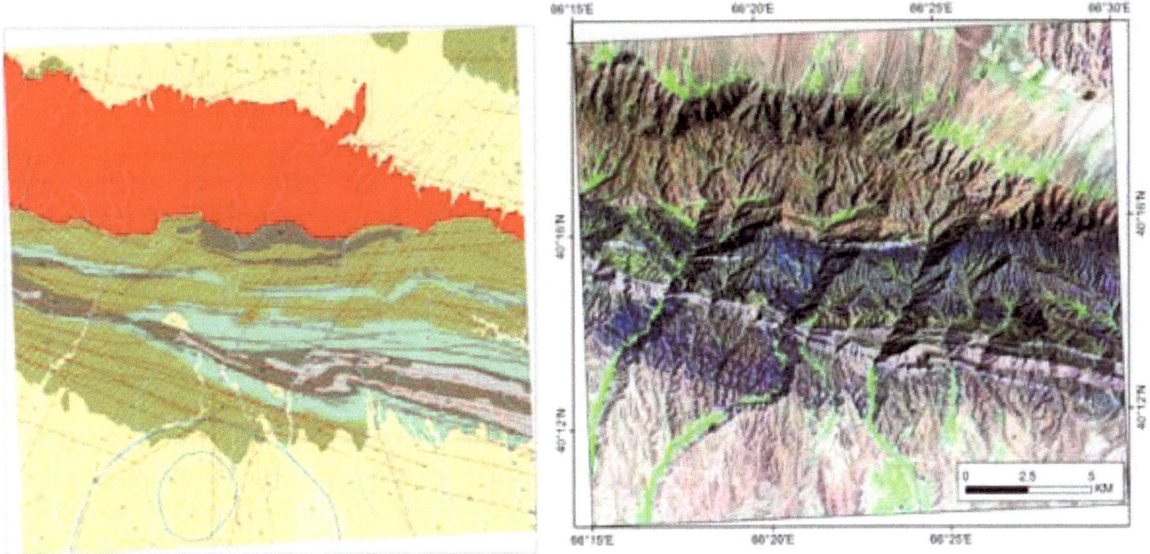

Figure 3.1.11. Terrigen-carbonate-shale SDC (left) and space image (right) distributed in the Southern Nurota area

In the South-Nurota region, the described formations are exposed in ridges (chashma suite). Chashma suite is composed of sandstones, gravelites and shales. It is more than 1000 meters thick. The lower limits are unknown. According to R.R.Usmanov, it was re-covered with the layers of the gypsum suite. In space photos, the color of flowing-blue, flowing-brown is expressed in photos.

Metaterrigenous SDC, The complex is distributed in the southeastern and southwestern parts of the Southern Nurota uplands. All rocks are strongly metamorphosed. The contacts of these deposits with the adjacent formations are tectonic. The deposits of the suite extend in the form of two tracks with a width of 1.5-2.0 km - from the Majrum fault in the west to the small village of Charbog in the east. It was isolated for the first time by R.R. Usmonov in the village of Little-Chorbog and described as graptolites. Contacts with surrounding suites are tectonic. The suite consists of interbedded sandstones, siltstones, limestones and shales of varying thickness. In the composite image of the 7,4,1 (RGB) option in space pictures, the complex is observed with a change to a purple-violet color.

Terrigenous SDC, O_3. It was developed in the north-western part of the South-Nurota ridge (Fig. 3.1.12). In addition, in Gobduntov, they were exposed on the surface of the Ordovician almond and Llandoverian suites as part of the krut and jazbulok suites. Deposits are represented by mica-silica, siliceous, clay, chlorite-sericite, carbon-silica-mica shales, argillites, siltstones and sandstones. Basic tuffs and dolerite-basalts are found in the Jazbulok suite. In addition, according to V.V.Mikhaylov, they contain diabases, gabbro-diabases, and apodiabase porphyrites in the form of silt-like piles, dykes, and stocks.

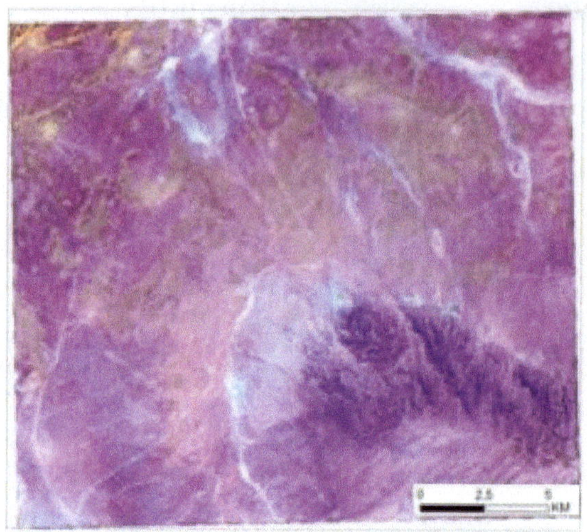

Figure 3.1.12. Existing terrigenous SDM in South Nurota area (left)
and space photo (right)

The deposits of the Bodomchali suite correspond to the Kalsara suite of Upper Cambrian-Lower Ordovician age. The Jazbulok suite is described in accordance with the deposits of the Lower Silurian Naukatsoi suite. In a composite image, the photoshoot is represented by a color gradient. The complex consists of siltstones, argillites, sandstones, clays, shales and gravelites. It has a ribbon-like structure, lying on the lowlands of the relief. 5=R, 4=G, 3=B variant image has brown feature.

Volcanogenic-terrigenous-shale SDC, S1. It developed in the Bitoksoi Basin in the South Nurota region - in the western end of the Kokcha ridge, in the Kara-Karga valley, in the Varadjon, Kurban region. It was isolated by P. N. Podkopaev and others in 1961 (R. S. Khan, 2006). It is represented by volcanic-terrigenous-shale SDC, consisting of limestone, sandstone, siltstone, gravelite, shales, tuffs, and composed of rhythmically (evenly) layered sandstone, siltstone, shale. The complex is colored from pale to dark dark gray. Its thickness is 400-500 m.
Stratigraphically, upper shales and siltstones are overlain by interbedded sandstones with thin layers of siltstones and shales. In the images of 5=R, 4=G, 3=B options, the complex has a dark-brown character, and in the cosmic pictures, the flow is distinguished by its dark phototone between the rocks.   Carbonate SDC, S-D. The complex was first mapped in 1986 by I.A. Pyanovskaya in the Vaush hills on the southern slopes of the Southern Nurota inter-ridge. The carbonate structural-deciphering complex consists of sandstones with mixtures of limestones, dolomites, siltstones, clays, gravel materials. Its thickness is 450 m. Contacts are tectonic. It is mainly distributed in the southern and southwestern parts of Southern Nurota. It occupies a high hypsometric position in the terrain. In the composite image, the photo ink is displayed in a liquid color.

Silicic - carbonate SDC, D-C. Separated by I.A. Pyanovskaya in 1986, according to the results of the AFGX works on the scale of 1:50000 on the Oktov

48

Mountains in the Southern Nurota Range, the Andin, Bakhiltov, Charkhansoy, Kokcha, and Zafarabad suites are included in this series (R.S. Khan, 2006). Lithologically, it consists of gray, white, various-grained, massively layered quartzites, marbles with lenses of sandstones and conglomerates, marbled limestones. It is more than 1000 meters thick.

Silicic-carbonate SDM is represented by limestone, marble, marbled limestone, from carbonate rocks with a silicic base. The photo color of the complex varies from pale to dark gray (Fig. 3.1.13).

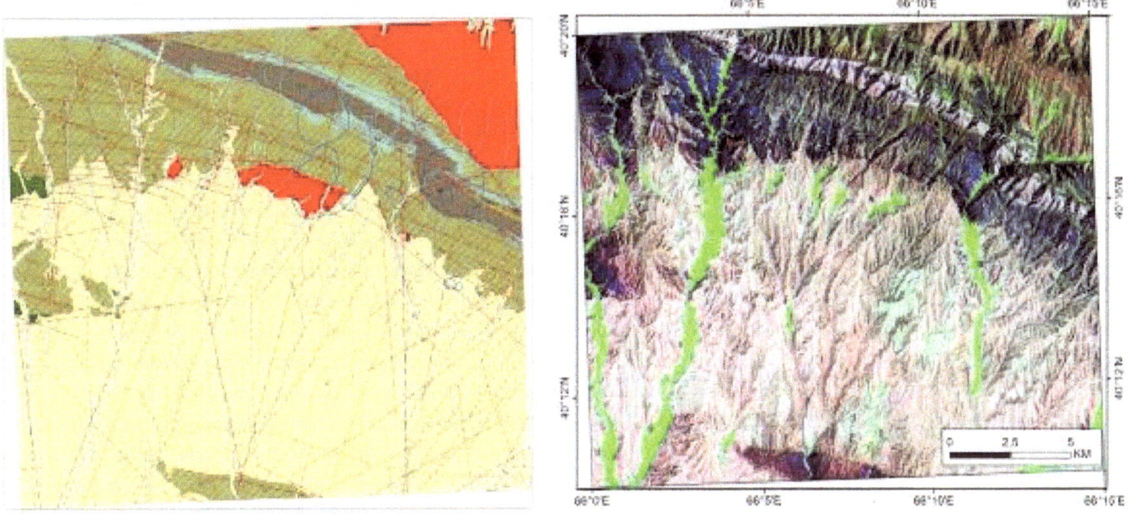

Figure 3.1.13. Silica-carbonate SDM (left) distributed in the Southern Nurota area and its appearance in space (right)

Volcanogenic-carbonate-terrigenous SDC, S2. The complex is distributed in the southwestern part of the Southern Nurota region and consists of marbles, gravelites, shales with lenses of conglomerates, sandstones, limestones. Occupies small areas. Space images show the complex in colors ranging from blue to white.

Terrigen SDC, K_2. This complex developed at the western ends of the Southern Nurota ridge in the Oktov, Karatov foothills, Kokcha and Karasigir hills.
The complex is composed of gray and greenish gray clays, sandstones, siltstones, conglomerates and gravelites with layers of sandstones. This complex has a gray to gray color in the cosmic image (Nurkhodzhaev, 2018). Carbonate-clay SDC, K_2. The complex developed in the western ends of the Southern Nurota ridge (in the foothills of the Oktov, Karatov mountains, in the Kokcha mountains, etc.). It is composed of carbonate-clay SDC, conglomerate-sandstones, limestones, greenish-gray clays and siltstones. In space photographs, the complex is depicted in shades of gray-green to white (Fig. 3.1.14).

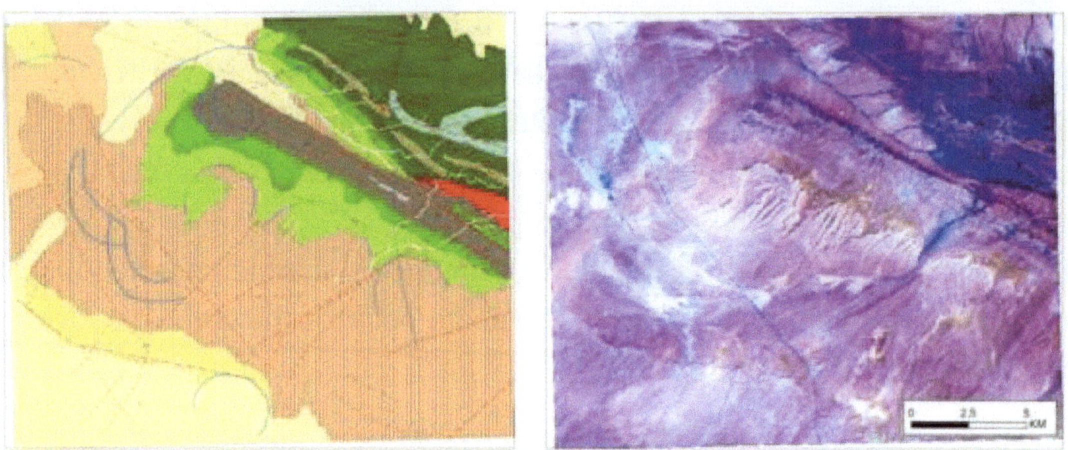

Figure 3.1.14. Carbonate-clay SDC distributed in South Nurota area (left) and space image (right)

Clay-carbonate SDC, 1-2. This SDC is mainly distributed in the southwestern and western parts of the Southern Nurota region. It is composed of clays, shale-limestones, dolomites, siltstones and gravelites. White, flowing gypsum, oolitic sandstone limestones and dolomites are identified in the section. Corresponding to various deposits of chalk, sometimes it lies with washing, and it is closed in accordance with the Bukhara suite (R.S. Khan, 2006). Its thickness is 200 m. This complex is deciphered in pink and dark gray in space pictures.

SDM with molasses, P_{2-3}. In the region, molasses SDM deposits are distributed in the western and southwestern parts of the Southern Nurota ridges. It consists of gray, gray, gray and brown marls with layers of limestone, sandstones, siltstones, and clays. Its thickness is 70 m. In space pictures, it is depicted in shades of blue to dark blue.

Clay-carbonate SDM, N_2. Clay-carbonate SDC is widely developed in the western and northern directions in the Southern Nurota area. It is composed of clays, gravelites, sandstones, siltstones and limestones. In space photos, this complex is separated by a light blue photo.

Clay-sandstone SDC, N_1. In the studied area, this complex is mainly located in the southwestern parts of the Southern Nurota uplands. The complex is represented by red clays, sandstones, siltstones, and limestones. In space pictures, it is depicted from blue to white.

Proluvial SDC, Q_{II-III}. This complex is spread over the entire area of the Southern Nurota uplands. The section of the complex is represented by loess rocks, sand rocks, supes, sandstones, gravels, conglomerates, pebbles, small gravels. The thickness is 100 m. Space images show the complex in shades of blue to green.

Alluvial SDC, Q_{IV}. In the region, the deposits of the complex are developed not only in the foothills, but also in all the riverbeds, ridges and terraces of the mountain formations (Nurota and Molguzar mountains). In the cross-section of the complex, gravels with layers and lenses of clays, rounded large stones, sands, small gravels, supes, clays are observed (R.S. Khan, 2006). The thickness is up to 50m.

Large parts of the area are occupied by alluvial SDM, QIV (sandy soils, gravels, small gravels) deposits. This complex has different colors in the cosmic image. In processed space photos, green colors are distinguished significantly, and natural colors are distinguished by dark colors. It is widespread mainly in the north-west and south-east parts of the area.

3.2. Field Control Inspection of Remote Sensing Materials of the Earth

Field inspection work is one of the main stages of decoding and serves to clarify the initial version of the remote basis, as well as to confirm the isolated cosmogeological objects. In cosmogeological research, field work is mainly carried out along with the passage of geological routes, description of geological observation points and other methods.

Deciphering of cosmogeological objects, intensity of linear and tectonic disturbances, relationships of structural-decoding complexes are determined by observation route studies (Togaev, 2018).

During the implementation of monitoring routes, field decoding is always carried out, the boundaries of decoding and the nature of structures are clarified in space photographs (SPh), (photographs of rock types and structural elements). The main purpose of this is to fill in the cosmogeological signs isolated during the initial deciphering work. Also, the signs of linear tectonic structures, color anomalies are distinguished by visual decoding.

Cosmogeological objects expressed in space imaging materials and separated in sph of different scales, including linear, field, etc., factors of exogenous and endogenous nature on the surface of the earth are connected with the influence of the complex. As a result of the initial decoding, a large number of tectonic structures, their intersection lines, color anomalies, etc. In this regard, field decoding works are mainly carried out with the aim of distinguishing signs of objects and signs of structural-tectonic and other causes of endogenous description.

Tectonic faults, annular structures, and SDC are the main structural elements in field monitoring. In connection with this, the main attention is paid to the above-mentioned structures during the field inspection. Decipherable signs of tectonic cracks are different: straight sections of valleys, relief elements, sharp changes in their directions, relief microforms, springs, linear formation of plants, clearly expressed in linear manifestations of exogenous processes. All of them, he points out, are mainly controlled by regional structures rather than local ones.

Tectonic movements can increase in the formation of annular or semi-annular relief and weakened zones. The separation of SDC depends on many factors, including the representation of these complexes, the dependence of the image on the scale, the presence of brightness with the SDC surrounding the complex, the degree of tectonic deformation, etc. Its boundaries play an important role in understanding these complexes. The fact that the territories of the Republic of Uzbekistan have been geologically studied at a high level, and the openings to the surface, as well as the fact that certain deposits are closed, are a very urgent problem in determining the need for additional studies and the prospects of closed areas. The methods of conducting searches in closed areas differ significantly from the methods used in open areas, first of all, research is carried out on a collective basis, using several methods. Field decoding of space photo materials was carried out in the following ways: - field observations in the research area; - checking the compatibility of the decryption object; - registration of decryption objects; - comparison of decoding objects in space pictures; - description of geological observation points; - creating a structural-lithological section, etc.

Such field-specific observation works (Chuya region, Karagaza pass, Langar region, Kokbulok, Western Zhailov, Dara region, Oktov region, on the northern slopes of Pashattov mountains, Kyzilkuduq, Urajon, Kokcha, etc.) in the western and eastern parts of the Southern Nurota region. went. Below are the results of the work carried out in some fields on field geological routes. As a result of geological observation, geologically interesting zones were observed on the slopes of the Chuya section. As a result of sampling and field interpretation of space photographs, tectonic disturbances in latitudinal and meridional stretches along the contacts with carbonate, terrigenous, terrigenous and intrusive rocks were confirmed and clearly distinguished against the background of space photographs (Fig. 3.2.1).

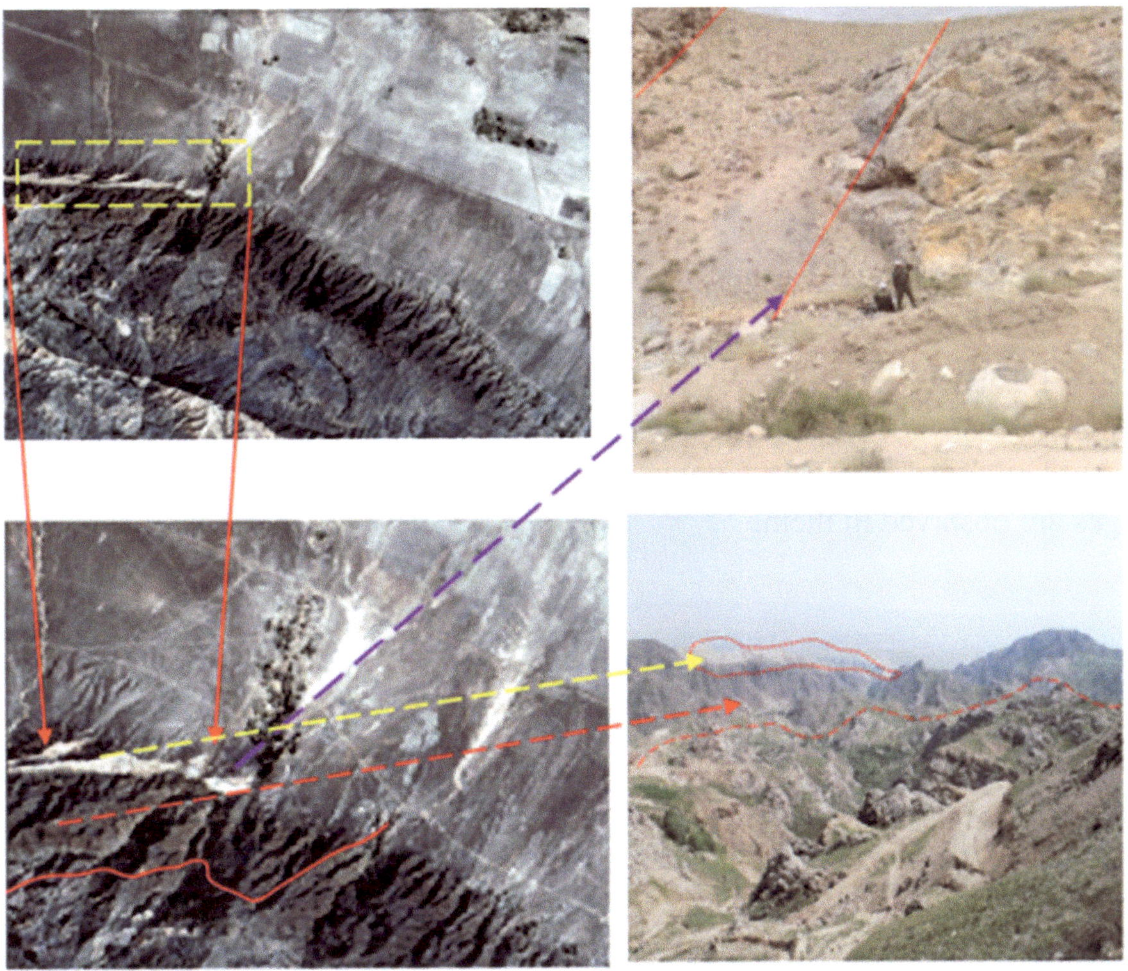

Figure 3.2.1. The situation of comparing the image taken in the field with the space image (South Nurota region)

Meridional faults pass through the granitoids, forming saddle-like folds and deep gullies at the watersheds of the ridges. As a result of monitoring work, on the north-eastern slopes of the South Nurota region, 7 km south of the village of Chuya, in the contacts with carbonate deposits and Devonian deposits, the copper content is from 0.01 to 0.03%, tungsten is up to 0.1%, gold 1 ,0-1.3 g/t, silver up to 5 g/t. The above image shows the initial decipherment work and a space photograph compared with the field photography of this site. It can be seen from the space photo that carbonate rocks are painted in pale colors, terrigenous rocks are painted in pale pink, grayish colors, and intrusive rocks are depicted in pinkish gray.

Karagaza region. The tectonic disturbances, geologic-structural features, intersection nodes of photostructures, and changes in photoreflections established in the initial deciphering work were partially confirmed in the field control work.

Tectonic disturbances were mainly observed in granodiorites, soil closures or openings. The visible changes in them are poorly expressed. The contact of the marbles with the granodiorites appears to be tectonic without any major alteration.

Only at observation point No. 1005, terrigenous rocks were intensively altered, bleached, and limonitized.

Langar area. During the field inspection and confirmation work, the fault zone and lithological boundaries of rocks of different ages, as well as zones of rapid change of rocks between granodiorites, revealed in the initial decoding of space photographs.

Among the granodiorites (cut of the Chuya-Langar road) a zone of secondary rapid changes was distinguished. In addition, dykes (veins) of granite-aplite with a thickness of three to seven meters were identified. The zone goes through the watershed towards the village of Langar (Figure 3.2.2). In the south, in the basin of the Moika tributary, 2 outcrops of marbles have been identified, which form rocky ridges in latitudinal stretches. Contacts of marbles with granodiorites are tectonic, no changes are observed in them.

In the

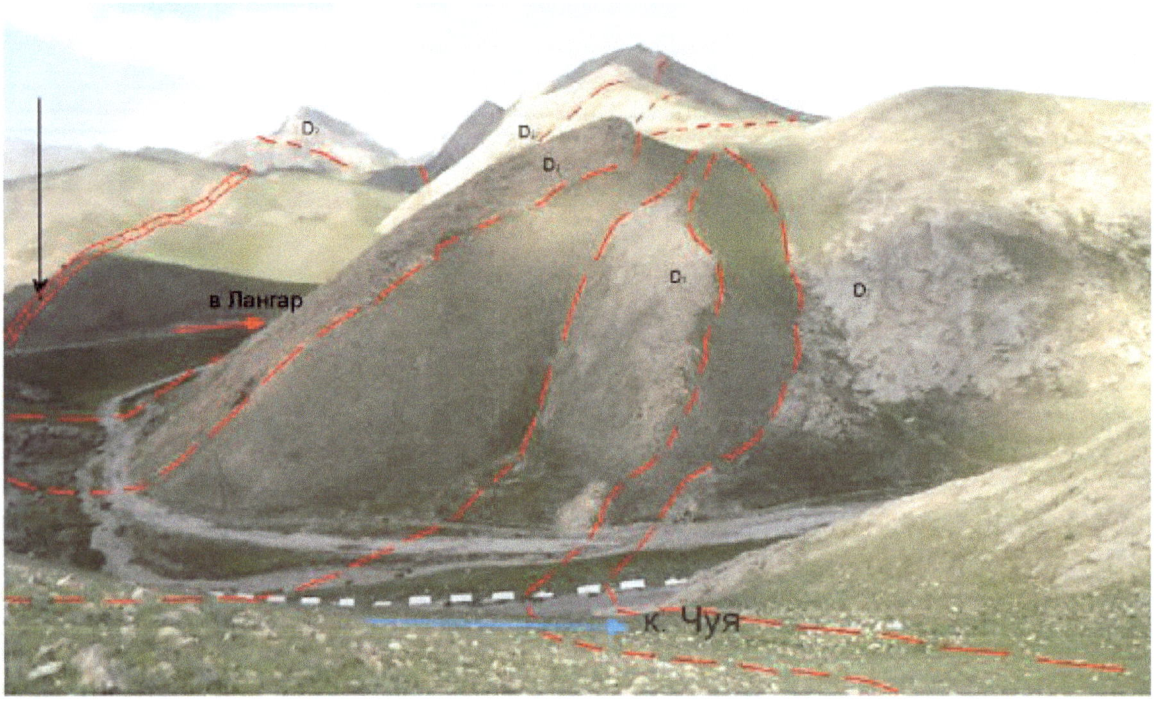

Figure 3.2.2. Gold mineralization zone in the Chuya section, north-eastern slopes of the South Nurota area

background of the space photo, the protrusions of the marbles are clearly noted, they form a flowing photoshoot. In space pictures, the photo of the disturbances is clearly expressed in the form of green stripes.

West Zhailov region. The contact of granodiorites with the Devonian marble deposits is recorded as a result of field control investigations. The contact was recorded on the spot and on the space photo. The contact is described as "cold" without any changes. The southern slopes of the Oktov Mountains were explored. In general, in the south-western part of the territory, on the northern slopes of Mount Oktov, geologically significant sections of interest were identified, except for observation points 1004-1006 and change zones at observation point 1011 (Fig.

3.2.3). Linear, point samples were taken for these points and the final analysis results were obtained by closely linking the description of objects in photoanomalies in space images with field observation work.

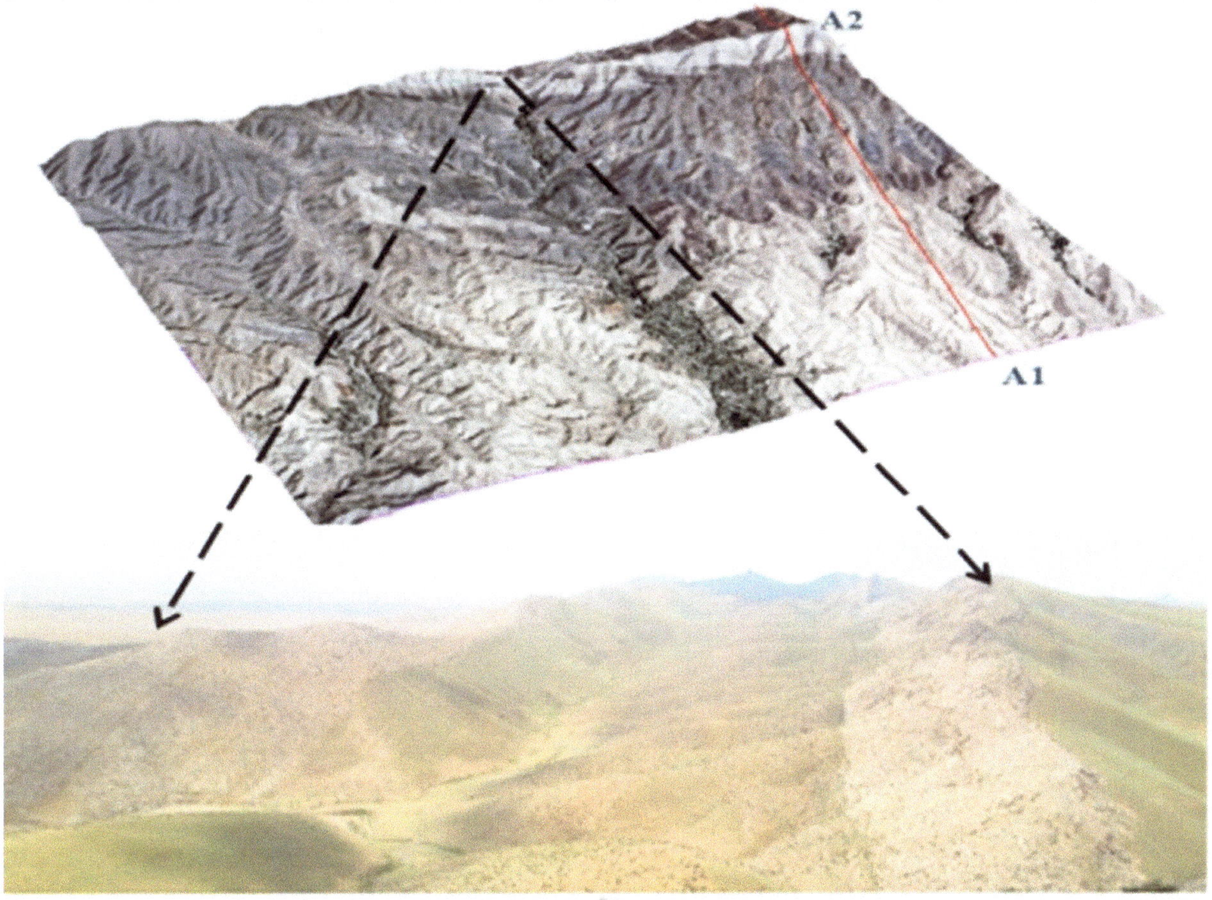

Figure 3.2.3. Comparison of decoded space image with field image (Oktov Mountains)

Gorge area. At this point, the openings of the exposed Carboniferous deposits between the intrusive and carbonate rocks of the Devonian age were mapped and monitored with the conducted field observation and inspection works. The discoveries of the Carboniferous age were recorded on the background of the space photo by its own photo. Photos of terrigenous rocks are painted in pinkish-gray color. Intrusive formations have a much darker pink color, and carbonate rocks are represented by a light gray color. The geological route was continued on the northeastern flank of the Oktov anticline. In space photographs, the flanks of the Oktov anticline are clearly distinguished by their light-gray photo. The anticlinal core is eroded and represented by granites. Latitudinal and NW-trending tectonic faults are observed, which are also well visible on the ground and in space photographs. Spot samples were taken along the alteration zones.

Aktov plot. Field interpretation of space images established a zone of sublatitude extensional tectonic faults cutting Devonian rocks. Tectonic contacts of Ordovician-Silurian and Devonian rocks were observed. A thick siliceous-carbonate layer of the Devonian period was mapped. The thick layer is characterized by massive and layered limestones with a fairly flat topography and distinct beds, forming thin layer formations. The Ordovician-Silurian rocks are represented by siltstone shales, ranging from turbidite to gray quartz, with small veins. It is observed that the thickness of Devonian carbonate rocks increases towards the west. In the background of the space photo, it is clearly noted in the off-white colors.

At observation point 1100 on the right bank of Maidonsoy tributary, a quartz-silicon vein was established in Devonian limestones. Quartz is non-mineral to the naked eye. To the west of observation point 1100, the contact of Ordovician-Silurian and Devonian rocks is established (Fig. 3.2.4). The contact is tectonic, strongly modified. Ironization and brecciation zones are observed in pre-contact rocks. Between observation points 1094-1096 ring-shaped structures were noted. This ring-shaped structure is clearly visible in the background of the space photo.

Quartz-mica shales with small veins of quartz were observed at observation points 1101-1104 along the field observation routes. In space pictures, shales are painted in shades of gray, dark gray. Along the streams, rift tectonic faults in meridional directions, which are cutting, were established. Azimuthal orientations of rocks are 20-250, orientation angles are 50-620. Along the course of the route, towards the west, the rocks gradually pass into silty shales with separate layers of coal shales. SDC are clearly recorded against the background of the space photo. Tectonic faults on the right bank of the Maidonsoi River have mainly a north-west direction (Fig. 3.2.5).

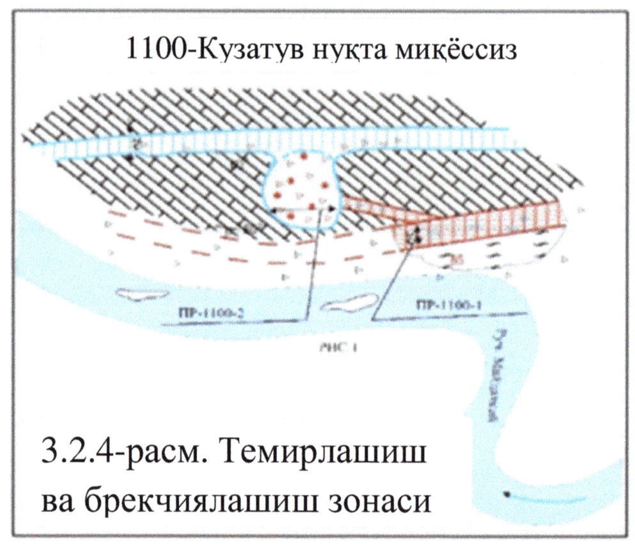

3.2.4-расм. Темирлашиш ва брекчиялашиш зонаси

Figure 3.2.5. Mineralized zone, right bank of Maidonsoy river

During field observation, tectonic faults were confirmed and interpreted in detail in space photographs. With the observation work carried out in the field, the zones of earth faults, the lithological composition of carbonate rocks were studied in the area, at the same time, the zones of change manifested in the initial decoding were determined. Excavations of carbonate rocks of the Devonian period were determined by cosmogeological research. Beds were clearly marked by the color of the discharge in the space image. In KS, carbonate rocks are colored pale gray. On the northern slopes of Mount Pashattov (12 km from the city of South Nurota), exposures of Devonian deposits were observed. In the space photo, these openings were clearly recorded in their leakage photototus (Fig. 3.2.6).

Also, during the observation route, intrusive formations with veins of small thickness of quartz, dark-red intrusive formations were detected (Figures 3.2.6-3.2.7). Against the background of KS, these intrusive rocks are distinguished by dark-gray phototext, SDMs are clearly marked. Carbonate rocks are widespread in the area and extend in the north-east and south-west directions, and are represented by limestones and marbles [82].

Figure 3.2.6. Granitoid SDM in the eastern part of the area

Figure 3.2.7. The zone of warping and quartzization

Kyzylkuduk region. Tectonic disturbances in sublatitude stretches were noted during decoding of the space image in field conditions. Outcrops of terrigenous rocks were identified. On the right side of the nameless stream, tectonic disturbances were revealed in the space image, and this information was confirmed during the investigation. In the background of the space photo, the openings of terrigenous rocks were noted in dark-gray colors. In the neighboring areas, Silurian, Cambrian, and Ordovician deposits were mapped with the field inspection work. Openings are recorded in a special bright color against the background of Sph. The photos of terrigenous rocks have a light-pink-gray color.

The main structural elements of the area were separated by summarizing the materials in the area with field decoding works. The cosmostructural features of the area, which are considered to be the main structural elements, are SDC, tectonic disturbances, meridional elongated earth faults of different degrees and annular structures. The field work was carried out with the help of a GPS navigator, connecting geological points (SDC, fault zones, tectonic structures, rift zones, etc.) and mainly with a description of geological observations, a map of geologically interesting positions and a photographic representation. At the end of the work, the information on the territory was reflected in the map of factual materials (Figure 3.2.8).

During the field work, attention was paid to the sections of intersection nodes of tectonic structures or annular structures, isolated as a result of deciphering color anomalies, summarizing the results of cosmogeological research with geological-geophysical data, and strong twisting zones between cracks. Geological routes were supplemented with sampling and analytical analysis in laboratory conditions (Table 3.2.1.).

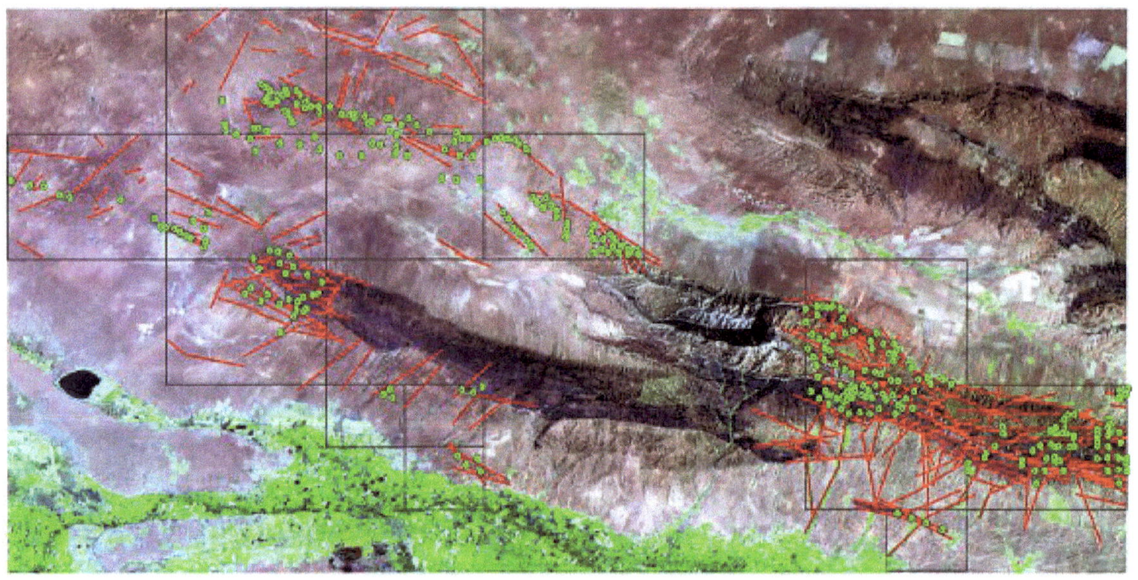

Figure 3.2.8 Evidence Map of Field Investigations

Conditional signs: ⊙ Observation points _______ Field decoding works

Table 3.2.1

Results of gold and silver assay (quantity g/t)

No	No example	Au g/t	Ag g/t	№№	№ example	Au g/t	Ag g/t
1	2	3	4	5	6	7	8
1	1005	1,5	≤1,0	38	1050	≤0,2	1,4
2	1007	1,3	≤1,0	39	1051	≤0,2	≤1,0
3	1011	≤0,2	≤1,0	40	1052	≤0,2	≤1,0

№				№			
4	SN-1-1	≤0,2	≤1,0	41	1053	≤0,2	≤1,0
5	SN -1-2	≤0,2	≤1,0	42	1054	≤0,2	≤1,0
6	SN -1-13	≤0,2	≤1,0	43	1055	≤0,2	≤1,0
7	SN -1-4	≤0,2	≤1,0	44	1059	≤0,2	≤1,0
8	SN -1-5	≤0,2	≤1,0	45	1059-1	≤0,2	≤1,0
9	SN -1-6	≤0,2	≤1,0	46	1061-1	≤0,2	3,6
10	SN -1-7	≤0,2	1,4	47	1061-2	≤0,2	≤1,0
11	SN -1-8	≤0,2	≤1,0	48	1061-3	≤0,2	8,2
12	SN -1-9	≤0,2	4,4	49	1062	≤0,2	≤1,0
13	SN -1-10	≤0,2	≤1,0	50	1069	≤0,2	≤1,0
14	SN -1-11	≤0,2	≤1,0	51	1070	≤0,2	≤1,0
15	SN -1-12	≤0,2	5,0	52	1071	≤0,2	≤2,2
16	SN -1-13	≤0,2	5,0	53	1071-2	≤0,2	≤1,0
17	SN -1-14	≤0,2	7,2	54	1072	≤0,4	≤1,0
18	SN -1-15	≤0,2	10,2	55	1072-1	≤0,2	1,6
19	SN -1-16	≤0,2	5,2	56	1073	≤0,2	≤1,0
20	SN -1-17	≤0,2	≤0,1	57	1083	≤0,2	≤1,0
21	SN -2-1	≤0,2	≤0,1	58	1087	≤0,2	≤1,0
22	SN -2-2	≤0,2	≤0,1	59	1100-1	≤0,2	≤1,0
23	SN -2-3	≤0,2	≤0,1	60	1100-2	≤0,2	≤1,0
24	SN -2-4	≤0,2	≤0,1	61	1333	≤0,2	≤5,0
25	SN -3-1	≤0,2	≤0,1	62	1334	≤0,2	≤5,0
26	SN -3-2	≤0,2	≤0,1	63	1337	≤0,2	≤5,0
27	SN -2-1	≤0,2	≤0,1	64	1352	≤0,2	≤5,0
28	1011-1	≤0,2	≤0,1	65	1353	≤1,2	≤5,0
29	1011-2	≤0,2	≤0,1				
30	1011-3	≤0,2	≤0,1				
31	1011-4	≤0,2	≤0,1				
32	1011-5	≤0,2	≤0,1				
33	1011-6	≤1,0	11,8				
34	1048	≤0,2	≤1,0				
35	1049-1	≤0,2	2,0				
36	1049-2	≤0,2	3,2				
37	1049-3	≤0,2	1,0				

Decoding results of field investigation conducted in the area are the results of field observations. At the same time, modern software products based on GAT technologies were used during the research (Figures 3.2.9-3.2.12).

Figure 3.2.9. Decoding result of space photo materials

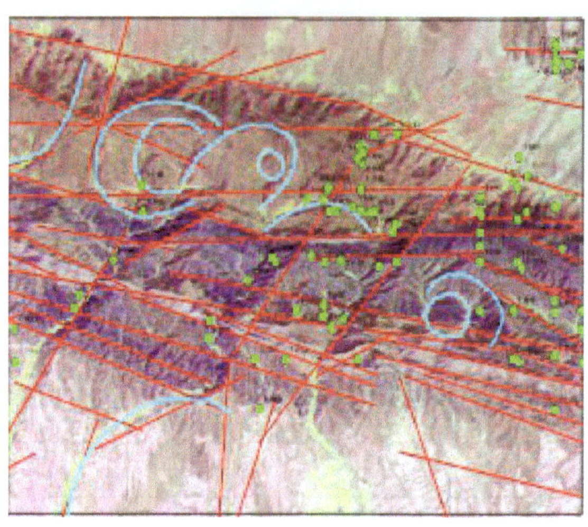

Figure 3.2.10. Geological observation points, linear and ring-shaped structures, the slope of Oktov mountain in the space photo

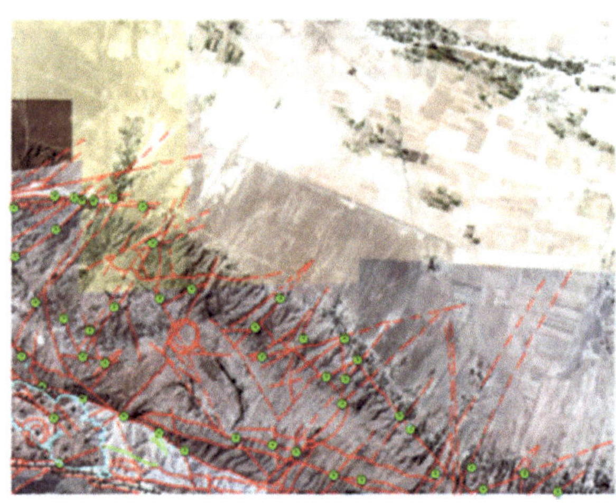

Figure 3.2.11. Observation points and the results of field decoding in the south-western part of the territory

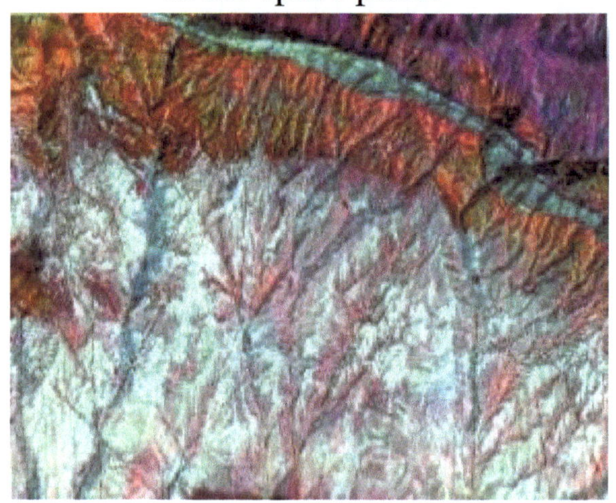

Figure 3.2.12. Observation points and results of field decoding in the north-western part of the territory

Conditional signs: 1-Field geological observation points, 2-linear structures, 3-annular structures.

Summary. Decoding of digital space images A large-scale digital remote sensing framework of the Southern Nurota region was created using geological, geophysical and geochemical data. As a result, completeness of geological information was achieved; Verification of the results of visual and automatic decoding in field observation work in creating a remote basis of the Southern Nurota region increased the level of reliability of the obtained data. As a result, the spread of ground cracks was confirmed in the observation decoding works; As a result of cosmogeological research, the geological and tectonic structures of the Southern Nurota area, granitoid, volcanogenic-carbonate-terrigenous, terrigenous, carbonate, clay-carbonate, siliceous-carbonate, mixtite structural-decoding complexes were distinguished. As a result, these structural-decoding complexes became important in monitoring the spread of tectonic earth faults and identifying thrust-type structures in digital space images in the area; The created remote basis will serve as one of the main cosmogeological maps for the regional analysis of endogenous mineralization in the Southern Nurota area and for planning future regional geological research. As a result, the remote basis created on the basis of the complex decoding of digital space images allowed to be filled with additional new information many times depending on the problems posed in solving geological problems.

CHAPTER IV: IMPORTANCE OF REMOTE SENSING MATERIALS OF THE EARTH IN REGIONAL GEOLOGICAL RESEARCH

4.1. Mapping Geological and Tectonic Structures in Space Images

The use of space imaging materials is one of the priority directions, which is the manifestation of cosmogeological objects integrated with natural objects on the Earth's surface - lineaments, arcuate and concentric circular photoanomalies. Such interest is explained by the fact that such photoobjects are often considered to be a projection of the deep structures of the lithosphere. In all regions of the world, these objects are of particular importance due to the fact that they control the location of many deposits of endogenous type, precisely with the degrees of manifestation of tectonic structures in space images (Nurkhodjaev, Shamsiev et al., 2017).

In determining the relationship between mineralization and magmatism in Central Asia, geologic surveys have been associated with linear and annular tectonic structures using aerospace methods (Glux et al., 2005).

In this regard, special attention is paid to the following in the structural-decoding complexes: - To evaluate the geological-structural informativeness of the materials of remote sensing of the Earth; - the reliability of the identified photostructures; - to the accuracy of the natural boundaries of limited photo objects. For this, there are methodological and practical approaches that take into account the variety of data (remote, geological, technological, natural, etc.) and allow creating a single strategy for processing space photographs in order to extract the main geological data (A.R. Asadov, J. Riskidinov, 2013).

The possibilities of structural-decoding complexes and tectonic structures depend, first of all, on the level of expression of reflected elements in the relief in the modern landscape. Naturally, scale plays an important role in this. Therefore, at the initial stage of deciphering, the main attention is focused on the description of the location and distribution of elements and relief components in the area, and on the forms of manifestation, as well as on choosing the most optimal scale for reflecting the natural changes in the form of geological-landscape objects in space photographs.

Research on the interpretation of space photo materials is aimed at determining the level of direct and indirect signs of structural-decoding complexes and structures and their manifestation in space photo materials.

Direct decoding symbols are pictures of photographs, photographs. They reflect the geometric and optical properties of objects. In practice, it is advisable to use not direct or indirect signs, but complex sign relationships (Nurkhodjaev et al., 2016, 2017, 2019).

It should be noted that in order to identify objects, for example, with spectral properties, it is not enough to know one spectral feature, it is also necessary to analyze the texture of the image. The texture of the image is determined to an important extent by the characteristics of geological structures, determined by the location of landscape components in latitudes (space).

The structure of the image is determined by the description of the shape, size and combination of landscape components, sometimes related to geological structures. In different areas of the photo, it is manifested in changes in the phototone, relief shape and texture. Therefore, the structure of the image is a stable sign. By changing the scale, the structure of the image changes significantly. Characteristic features of the structure are the reflected expression of the landscape image in which the landscape exists (Fig. 4.1.1). And in spectral signs - the types of unevenness on the Earth's surface, rocks and rock formations,

sizes and shapes; cracks; reflection It is important that it affects the color by determining the spectrum of reflected and absorbed light energy

Therefore, the role of direct and indirect signs according to the classification of signs is great in decoding. In the researched area, linear photostructures were noted in the decoding of remote photomaterials at different levels in different spectral ranges. They represent tectonic faults, microcracks, closed linear and fissure

Figure 4.1.1. Folding of layers

zones, landscape elements, straight line boundaries of two geological bodies clearly expressed in the modern relief. The form of expression of lineaments on the surface of the earth, which is considered as a determining criterion for their separation, is reflected in watersheds, river and stream valleys, bends in the relief, and differences in the color of photographs of their plots.

The analysis of linear structures showed that in all types of photomaterials, photostructures in the sublatitude and north-west direction and linear objects in the north-east direction are observed. As a result of rapid interpretation of these data and comparison with other data, mapping of geological structure and tectonic faults leads to an increase in the quality level of geological information.

4.2. Characteristics of Automatically Detecting Highly Prospective Mineral Positions

It is the manifestation of the signs of an event or process in the cosmic image in different spectral channels. First, the preprocessing of the space image or the processed space image is chosen, because the direct application of brightness characteristics of the original space image materials in solving many geological tasks does not always give satisfactory results.

To improve the results, the resolution of the space image is directly increased, the variability of the pixel value is reduced, which is achieved by normalizing the solar radiation by reproducing the spectral brightness, that is, the atmospheric correction is carried out, all these steps are calculated in the algorithms built into the GAT program (Busygin et al., 2006). Due to the fully automatic dependence of the interpretation processes of large volumes of data, the algorithms used for editing are calculated by averaging indicators. Clouding is a common problem in using space imagery. A number of software packages created for the processing of Earth remote sensing materials can reduce the effect of cloudiness in the image, but in such cases, additional information about the state of the atmosphere is needed during imaging (Busygin, 2005). To modify and edit the model under a clear image, the following basic steps should be performed: data acquisition, data analysis. Image input to the model must be a single, multi-channel file (channels 1, 2, 3, 4, 5, 7 combined using the Layer Stack command). The result of the work is an atmospheric edited photo, the file format is unsigned 8 bit. In the obtained KS, the benchmark mine or object is located in a cell (pixel) consisting of three layers, namely R1, G2 and B3 layers of the spectral channel. Each layer has its own attribute value, which serves to unify and describe the class, category, group belonging to the cell, or for the task of quantitative description of the properties describing this R1 layer 130, second G 132, third B 170 raster.

When changing the unit of space image channels, the values of R4G5B6 pixels will have different numerical values (R4 channel = 161, G5 channel = 178, B6 channel = 142). As a result, we get 6-valued pixels for one standard (132,135,171,178,142). At the next stage, it is required to solve two tasks, the first is to find the average numerical value and the second is to cover neighboring pixels at the same time, that is, to find the given information value by neighboring zones. The evaluation of the values of the surfaces between these points is carried out by averaging the values of the points of the neighboring zones, taking into account the degree of their proximity to the given point, but it is necessary to distinguish between the small and large zones that enter the ground of the reference object (Shamsiev, 2015, Glukh, Potorjinsky, Eisfeld, 2009). The purpose of displaying these zones is to study neighboring values and how they are distributed. For example, the standard object may be in the zone of small values, which may not be taken into account during processing.

The points corresponding to each standard cell with high and low priority values are processed to display the calculated rasters, that is, the total plot of different pixels is calculated. In the next step, the method of returnable residual distances (RDD) is used. In this method of interpolation, each input point has a decreasing effect with distance. The closer the point is to the cell being processed, the greater its weight. The clustering module (The Cluster/Outlier Analysis) allows to calculate the degree of difference with neighboring fields, weight values for the purpose of finding benchmark locations and apply values for their averaging (Fig. 4.2.1).

Space imagery uses distinguishing features such as shape, object boundaries, color, and pixel brightness to locate and identify objects. A multispectral Landsat 7 TM space image was used for the decoding of the space image.

Figure 4.2.1. Significant values of gold (Au) benchmark objects

The image is 8-bit from 7 channels, and three main components of selected colors are entered: R (red), G (green) and B (blue). The brightness of each channel fluctuates between 0 and 255, that is: [0; 255] is described in intervals. Space images are composed of arrays of discrete numbers, i.e. matrices, each element of which is three:

R, G, and B are value terms, so they can be expressed as some function. The task of interpreting space images is complicated by the variable features of many visible objects. The photos taken depend on the date of shooting, for example, the growing season, changes in river beds, snow cover, forest fires, etc. includes variable data in its content. In primary images, primary objects can be distinguished, which have such properties as stability, invariance in different processing, specificity, differentiation from other objects. Secondary, that is, objects that are not visible to the human eye and have their own spectral information, are separated using automatic methods (primary and thematic processing, classification, segmentation, methods of applying various filters, etc.) (Nurkhodjaev, 2017, Tokareva, 2010).

A new algorithm can be used to improve the existing approach of evaluating the correspondence between the states of objects using the maximization of the mutual information function of the space image and the benchmark. The idea of the new method is based on the simultaneous use of several types of space photographs and cartographic materials. Such an approach allows to impose additional restrictions on the solution of the task due to taking into account the geometrical and latitude relations between the space image and the reference image.

In connection with this, it is solved by a general algorithmic scheme for identifying prospective mineral objects and predictive-prospective plots, which are concentrated in the multi-stage processing of space images. In the complex algorithm of automatic representation of high-prospect objects, each algorithm solves a part of the general task, the quality of assessment decreases when one of the factors of the result analysis is lost when one type of task is not solved.

For example, in the Landsat TM space image, channel 1 has a range of 0.45-0.515 blue, channel 3 has a range of 0.63-0.69 red, channel 2 has a range of 0.525-0.605 green, and channel 6 has a range of 10.40-12.5 wave infrared (heat), channel 7 2.09-2.35 mid (short wave) infrared, corresponds to the absorption bands of radiation from hydrominerals (clay shales, some oxides and sulfates), which makes them appear dark and map useful service. Pixel processing of this image is related to the search for pixels and their regularities based on an algorithm (Tokareva, 2010). Pixel processing of digital space images according to this algorithm allows the values to be automatically classified according to their brightness differences, because changes in pixel values are associated with changes in geological structural coverage on the surface of the earth and are not taken into account in other similar methods. When using points of mineralization as reference objects of maximum (gold-bearing mineralization) and others, one of the signs of the manifestation of mineral deposits is the intersection of certain earth faults with mineralized positions (Fig. 4.2.2).

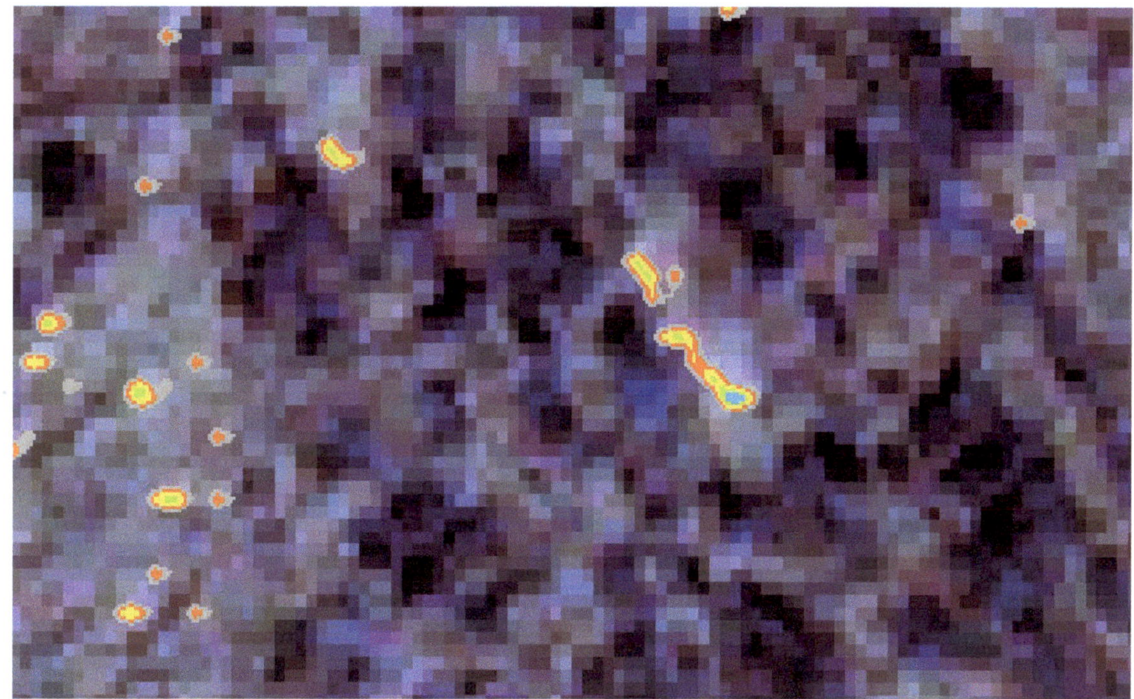

Figure 4.2.2. Halos obtained by statistical averaging

Most of the known standard objects are associated with linear and ring structures, their dispersion and high value structures. In order to reveal mineral positions, the results are filled and clarified using the algorithms determined by the mineralogical indices of the ASTER satellite image.

Until now, there are various methods of decoding, and one of them is the method of ratio and combination of channels. The number of combinations of channels on ASTER space image is much more than that of Landsat TM space images. In terms of the ratio of spectral channels (split, readout, addition, etc.) ASTER has a high advantage over other images, it is especially important that it shows signatures. Many methods are focused on the determination of one type of mineral, and several types of channels are used for one mineral (A.K. Glux, 1987). It is important to note that ASTER's spectral channel solutions range from 15 to 90 meters, channels 1-3 (VNIR) to 15 meters, 4-9 (SWIR) to 30 meters, 10-14 (TIR) to 90 meters. The complex functional algorithm is divided into 3 groups according to the spectral solutions, after the process, the results of the 3 groups are combined into the final result. Initial space image processing, comparison MNF, PCA methods were carried out for comparative evaluation of space images using automatic methods (Fig. 4.2.3).

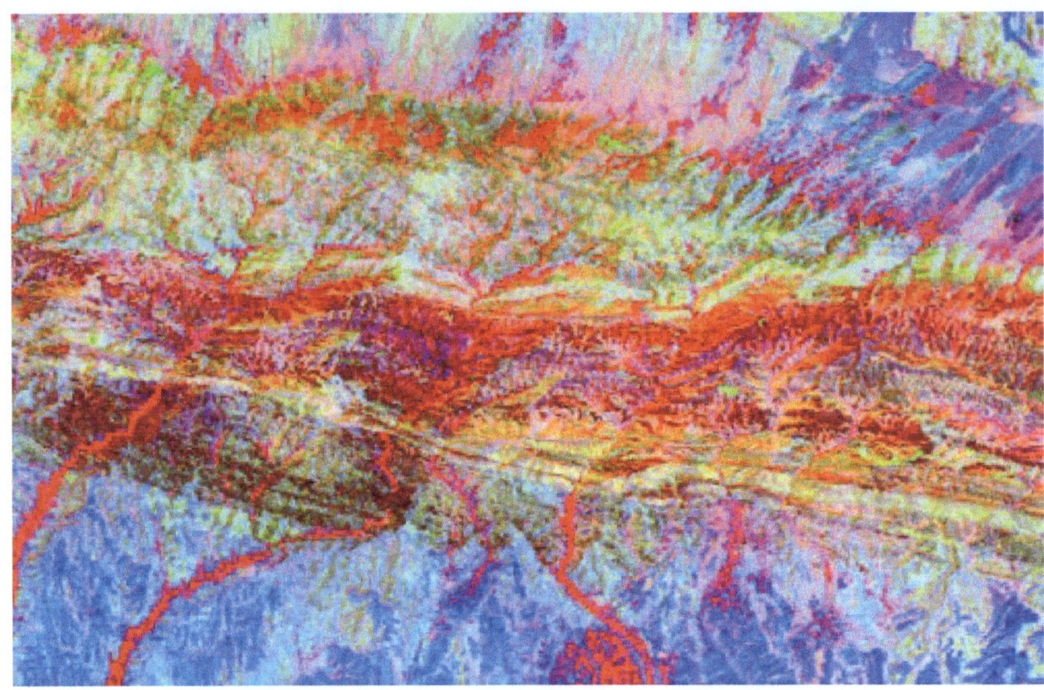

Figure 4.2.3. A space image obtained using the MNF function

The pre-processing in the above-mentioned methods gives important results, which can be applied to other spectral channel scaling methods. The results of the comparative comparison showed that the two signatures obtained in different regions of the Earth are completely identical, which made it possible to use other signatures in automatic search. The next step is to create a combined image from the images with a spectral resolution. This operation is performed using the Erdas Imagine program module "Layerstack...". When finding the mean and cumulative average values of the composite image, they are grouped and combined into a single chain of halos depending on their importance. ASTER SPH with 14 spectral channels can obtain more indices than Landsat when changing channel positions. Using 6 spectral channels with a resolution of 30 meters in the SWIR frequency, it is possible to allocate indices according to the known ratio algorithm proposed by the Japanese manufacturer Ninomiya (2003). The following data will clarify the intersection points by separating the prediction plots (4.2.4-4.2.5). This information is based on studies of mining areas and geological data.

In addition, it should be noted that the use of ASTER space images is mainly due to the possibilities of using combined channels, MNF, PCA methods have a high efficiency in interpretation. It also compares the data of this signature with the signatures obtained from space images, and when comparing them, their similarities are observed to a large extent. In this case, the possibilities of enrichment with different methods of interpretation increase.

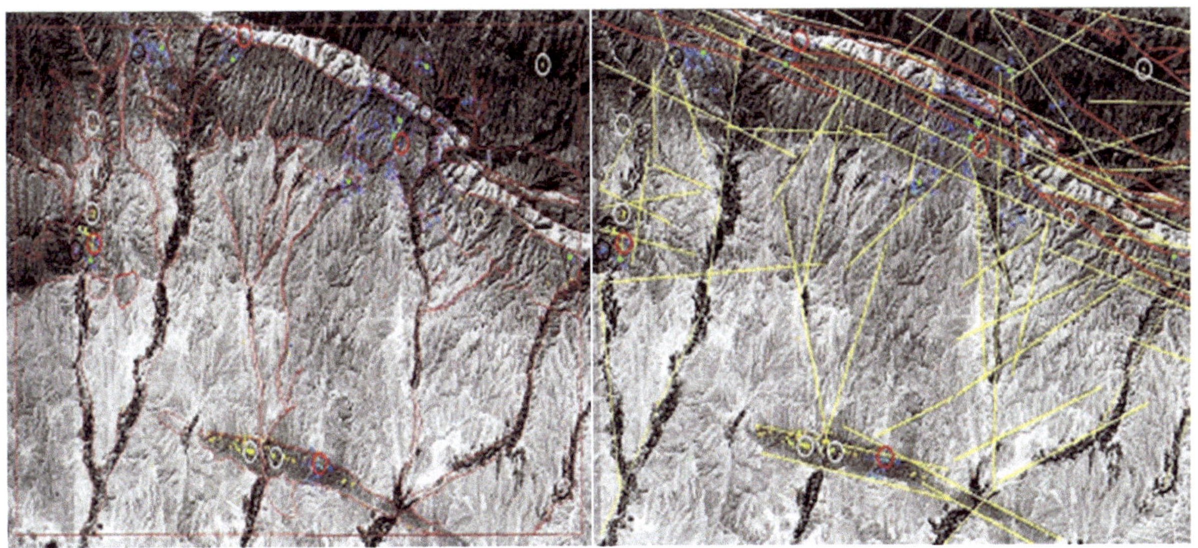

Figure 4.2.4. Linear structures and positions where the ore can be manifested.

Figure 4.2.5. Manifestation of zones of ore position in space image

Thus, the developed complex algorithm for decoding ASTER and Landsat space images allows using several types of algorithms and identifying various objects and their signatures.

It all depends on the knowledge and experience of the geologist when considering the complex algorithm and getting the final result. Analysis and assessment are also carried out with the help of cartographic data and field control verification work.

From these pictures, it can be seen that the areas defined by reference objects have a characteristic line, that is, they are partially continuous or separated depending on the accumulation of ore positions relative to the location of the reference points. The presented plots are used to predict the research area and it allows to determine the level of reliability of the results and to determine the applicability of the procedures used (Figures 4.2.6-4.2.8).

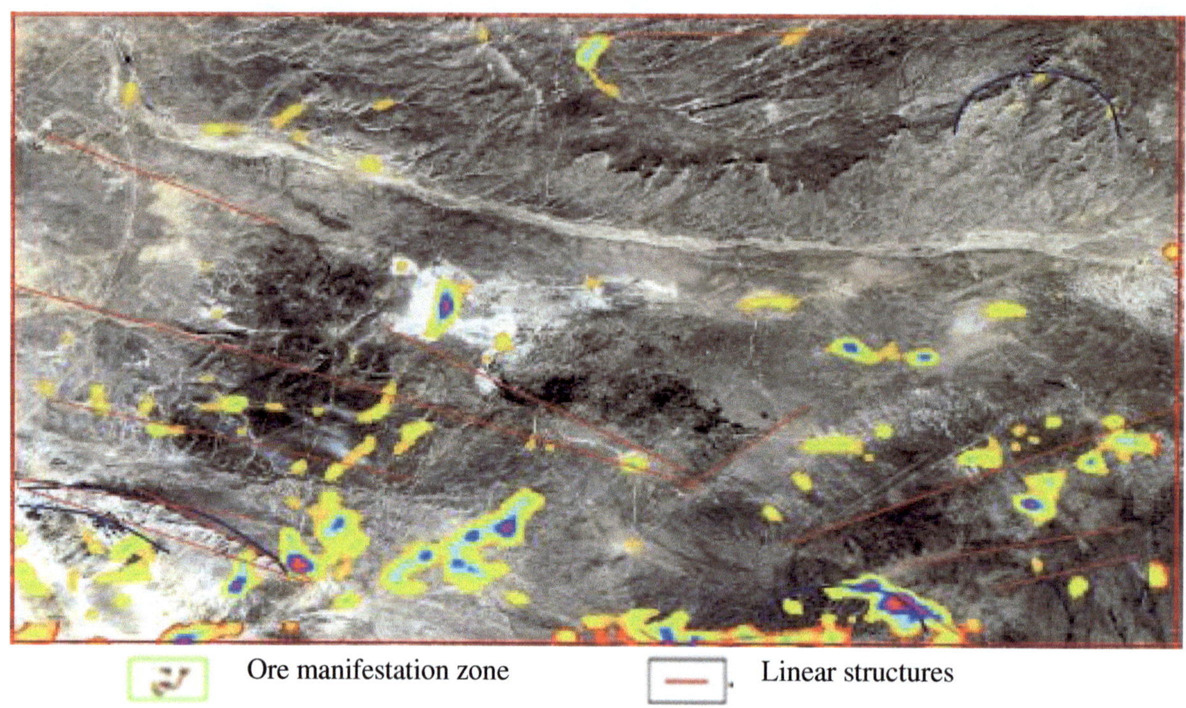

Figure 4.2.6. Spatial appearance of ore zones in the north-eastern part of the Southern Nurota region

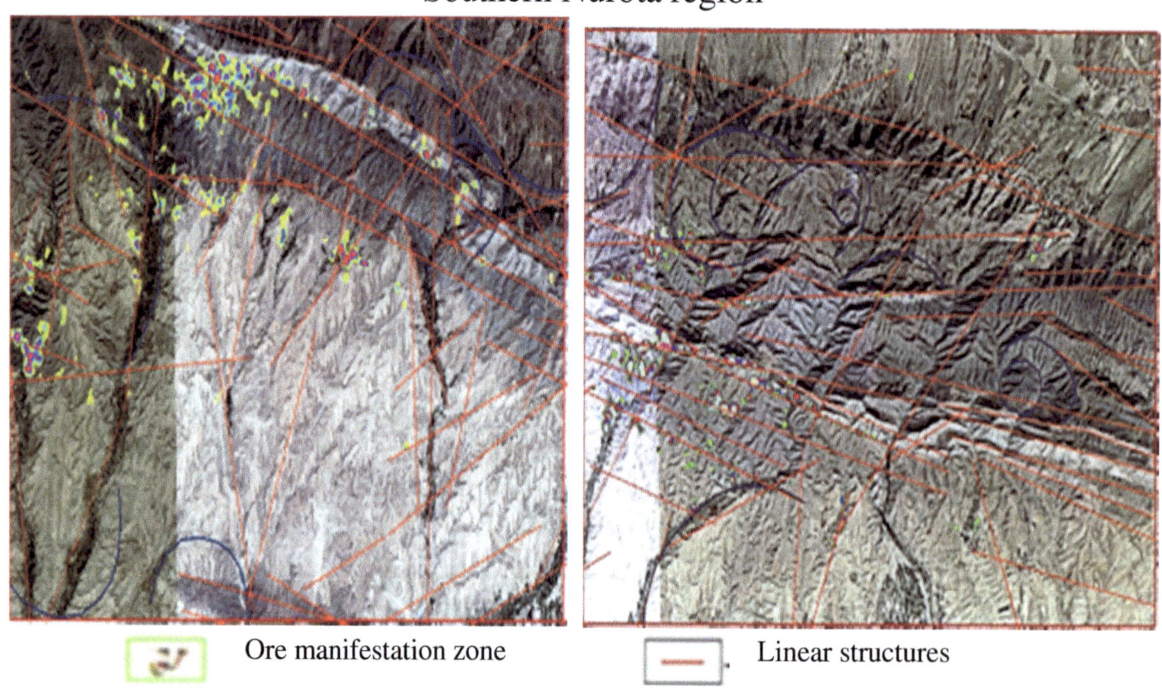

Figure 4.2.7. Manifestation of ore zones located in the north-western part of the region, on the southern slope of Mount Oktov

Figure 4.2.8. Manifestation of ore zones in the north-eastern part of the region in space images

4.3. Identification of Highly Prospective Areas and Positions

The main mineral resources in the Southern Nurota area are gold, silver, etc., therefore, the allocation of prospective areas is mainly based on the factors controlling gold mineralization.

The gold composition of the research area has been extensively characterized in modern erosion shearing, which was manifested in the conditions of the absence of exposed industrial deposits and the wide development of gold-bearing manifestations (Koloskova, 2005).

Prediction of gold mineralization is one of the most complex areas in geological exploration. In order to solve this problem, it is important to note the factors and signs of mineral control in the formation of gold mining areas. Mineral controllers include the factors that determine the location and formation patterns of minerals in mineralized regions and areas with specific mineral content, that is, they are mainly considered indicators of the formation of mineralization. The main mineral control that determines the patterns of location and formation of minerals is: lithological-stratigraphic, structural and magmatic factors (Koloskova, 2005).

Lithological-stratigraphic factors, the regular distribution of minerals in sedimentary and sedimentary-volcanic rocks of a certain composition and age. The location of gold deposits, ore occurrences and mineralized points in the Southern Nurota region has a clear stratigraphic control - it is primarily observed in deposits of Cambrian, Ordovician and Silurian age. During the processes of regional metamorphism, conditions are created for their subsequent deposition in the migration of elements, their enrichment, contacts, metamorphogenic quartz veins, subconformity layering of rocks. Silurian deposits are important ore-formers, at the same time, high metal content of coal-shale layers has been determined (Z.M. Abduazimova et al., 2004).

The following lithological features of the metamorphosed sediment thick layer are considered favorable for ore deposits: high permeability of the rocks, brightness of the section and rhythmicity of the gradual occurrence of individual facies. The high absorption of rocks is the result of specific physico-mechanical properties (for example, high degree of fineness (absorption, elasticity). Sands, limestones, metavolcanogenic and siliceous rocks are characterized by high brittleness and permeability (Koloskova, 2005).

Structural ore-controlling factors are fold and fracture and their associated structural elements that influence the location of deposits and ore bodies.

Structural factors. It is the main ore control in the locations of gold mineralization manifestations in South Nurota. Regional and local ones are distinguished among them, and they are important as ore controls of fold and crack structures. Depending on the importance of controlling ore, the fracturing structures in the area are divided into ore-generating, ore-transferring and ore-distributing, ore-forming structures (Sidirova, 2018, Koloskova, 2005).

The mineralizing structures controlling the location of gold mineralization in South Nurota are served by the Karatov and South Nurota thrust zones.

The ore-conducting structures in South Nurota are mainly:

North-east, north-northeast, north-northwest faults and transverse high fault zones are considered. They determined that gold mineralization manifestations belong to the structures in these directions in the intersection sections of the crush zone. Ore-distributing structures usually include block-forming structures and structures characterized by a much higher composition. Ore-conducting and ore-distributing structures are of particular importance in the location of gold-bearing occurrences in Southern Nurota. These structures are represented by fault faults in the north-northwest, sublatitudinal and northeast directions, with a length of 2 to 5-7 km and more, and are intersected by a rift zone, and make them a complex internal block structure, intersecting nodes of faults in different directions. forms.

Mineral-controlling structures mainly determine the physico-chemical conditions of ore deposits and mineral-enclosing horizons (zones, blocks) and the morphology of ore bodies trapped in them (Fedorchuk, 1991). In Southern Nurota, they are characterized by quartz vein and vein mineralization, bleaching, graphitization and limonitization of rocks.

Ore-controlling structures with gold mineralization include: 1) Intersecting, in different directions, unstretched, cracking and brecciation zones of rocks; 2) Cracking and crushing zones belonging to the layers of brittle (quickly brittle) rocks, corresponding to the layering of rocks; 3) Short cracks along which quartz veins and vein-like quartzizations are identified; 4) layering of separate layers formed as a result of plastic deformation of rocks in the locked parts of folds, thick layers in flexural bends, in isoclinal fold sections, along with vein-vein quartzization; 5) Less open ring-shaped and other small structures of tectonic faults with a length of 5-6 km.

Igneous ore controlling factors are defined as properties of igneous rocks that represent the manifestation of mineralization. The connection of minerals with igneous rocks is shown in: 1) genetic connection with certain intrusive arrays; 2)

72

paragenetic connection with igneous rocks or certain types of processes. A paragenetic correlation with the S_3-R_1 granitoid magmatism of the Shorok intrusive complex has been established for post-magmatic hydrothermal gold mineralization in South Nurota. The importance of magmatic factors as the main ore control in the location of gold mineralization is concentrated in the ore-generating role of magmatic processes. Gold ore mineralization is controlled by the intrusive position over the undrained magmatic bodies in the latitudes, their positions in the exocontact parts, the dike fields and the distribution of the belts.

Mineralogical signs. According to the results of direct mineralogical prospecting, high gold contents (more than 1 g/t.), quartz-vein formations with slime halos of pure gold and vein-vein quartz-sulphide mineralization were observed (Koloskova, 2005). Such signs belong to the fringe zones in the study area.

Indirect search signs in the Southern Nurota area include: 1) metasomatic changes of the rocks of the beryzite formation, recording mineral metasomatites; 2) vein and vein-vascular distribution of quartz mineralization; 3) slime aureoles of gold, scheelite, cassiterite and monazite; 4) exocontact-specific re-formations of surrounding rocks - hornification, amphiboleization of rocks, development of shale and increased contents of magnetite and pyrite; 5) is the presence of accelerated limonitization as a result of exogenous oxidation of sulfide minerals: pyrite, arsenopyrite, chalcopyrite in rocks and veins.

Geochemical signs. Direct geochemical tracers of gold-bearing mineralization include: 1) points with a high content of gold (Au one hundredth g/t.si and more); 2) secondary spread oreos of gold and marzipan; 3) high geochemical brightness of ore-forming elements; 4) regular structures of geochemical fields. Indirect geochemical search signs are the co-occurring aureoles of elements-trees of gold mineralization - As, Ag, Pb, Cu, Cd, Zn, and elements widely distributed on the surface of the ore - Sb, Hg, J, Cl, etc. In the regional plan, the latitudinal connections of the geochemical structures with the Karatov and South Nurota subduction zones have been determined (Koloskova, 2005).

Taking into account the above information, several high-potential - promising areas were allocated in the research area based on the results of cosmogeological research. They were previously separated by prospects based on the summation of geological-geophysical, geochemical data and the analysis of the formation factors of mineralization. In addition, 2 high-prospect areas were sampled, geological routes and field observations were conducted to confirm field observations. S.M. Koloskova's materials were used in the detailed analysis to distinguish high potential - promising areas. I. High potential - promising area. It is located in the eastern parts of the research area, in the southwestern part of the area, in the mountainous parts of the northern slope of the Oktov mountain, near the village of Chuya. The field consists of siliceous-carbonate deposits of D-S, volcanogenic-terrigenous shales of S1, granitoids of C2-3 and olistrostromes with siliceous, micaceous-quartz shales, dolomites and granitoids, structural-deciphering complexes represented by limestones, marbles, marbled limestones.

The field is controlled at the intersection nodes of sublatitude and northeast structures. In the course of field work, geological routes, approval points, and structural-lithological cuts were studied. 3 tectonic zones with large thicknesses and latitudinal directions were observed in the areas of the high potential field. The zones are characterized by the widespread development of limonitization, graphitization, and vein-veined quartzization. The zone of mineralization refers to the contacts of terrigenous rocks of the Carboniferous with Devonian limestones. In the zone of mineralization, the rocks are strongly crushed, limonitized, and cut by a large number of quartz veins. 7 km from the village of Chuya. in the south, in the tectonic contact (thrust) position of carbon with Devonian deposits, the content of copper is 0.01 to 0.03%, tungsten is 0.1%, gold is 1.0-1.3g/t, silver is 4.6 meters thick. up to 5 g/t (Fig. 4.3.1).

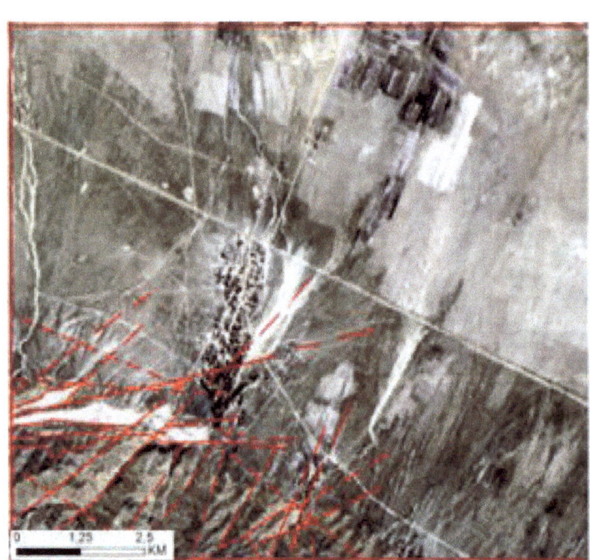

Figure 4.3.1. Decoding results for the southwestern part of the area (left) and distance basis (right). Conditional symbols are in Figure 4.3.2.

II. High-potential field. It is located in the eastern and north-western parts of the research area, in the mountainous parts of the southern slope of Oktov Mountain, near the Maidonsoy tributary basin.

The research area consists of Cambrian (€) terrigenous-carbonate-shales, SDC volcanogenic-terrigenous-shales of Silurian (S1) age, silicic-carbonate (D-S), granitoid structural-deciphering complexes, molasses SDC of R2, silicic and granitoids composed of structural-decoding complexes represented by argillites, clays, sandstones, limestones, shales, siltstones, gravelites, tuffs, marbleized limestones. The area is characterized by the wide development of tectonic structures, especially the distribution of shear-thrust zones, as well as sublatitude, north-west, north-east and meridional extension (Figure 4.3.2).

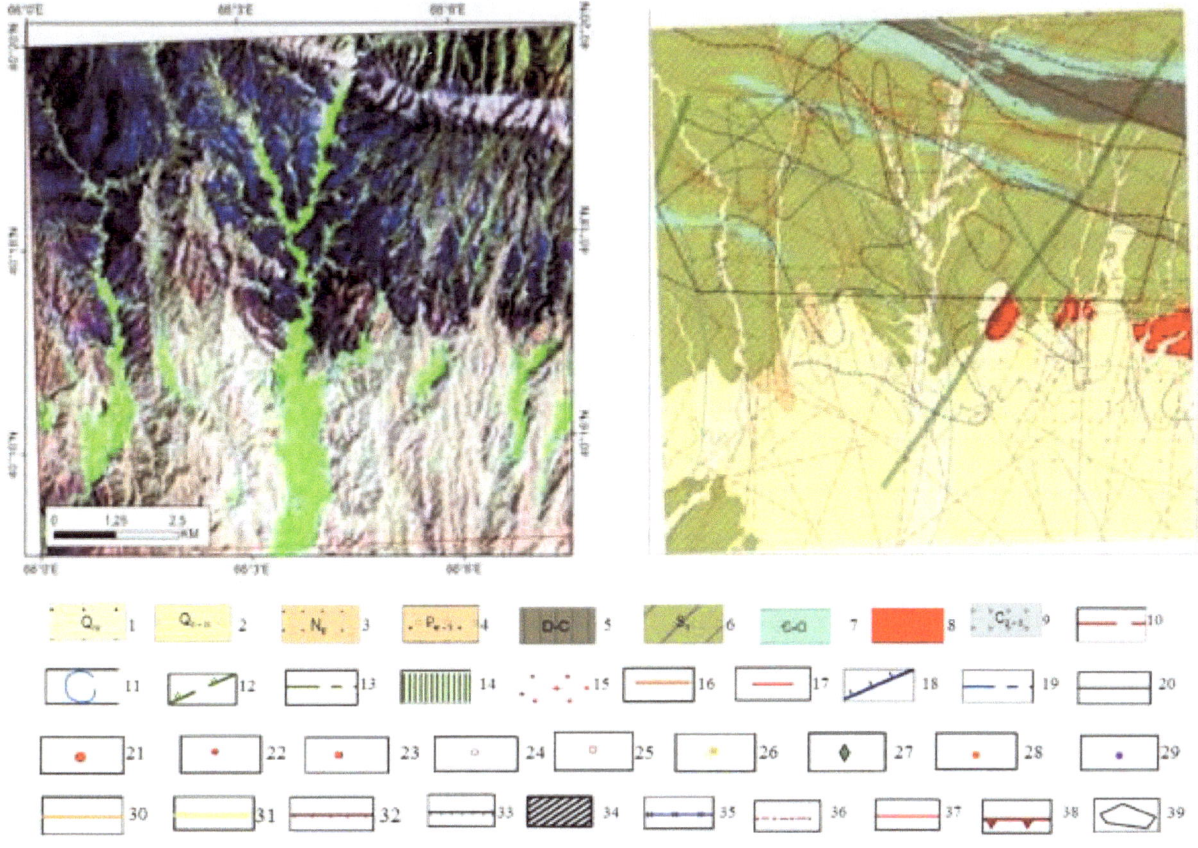

Figure 4.3.2. Decoding results on the north-west of the region, on the southern slope of Mount Oktov (left) and the result of the remote basis (right)

The field is controlled by thrust, as well as sublatitudinal, north-west, north-east structures junction nodes, where gold and margium mineralization points, gold, scheelite, monocyte slimy aureoles and local positive anomalies and negative anomalies of the natural field are observed. During the confirmation work in the field, geological routes, approbation points, and structural-lithological cuttings were studied. At observation point 1100 on the right bank of Maidonsoi tributary, a quartz-silicon vein was established in Devonian limestones. The contact of Ordovician-Silurian and Devonian rocks was established to the west of observation point No. 1100. In this place there is a strongly altered contact of the rocks. A zone of ironization and brecciation is observed in the pre-contact parts of the rocks. In the obtained samples, the amount of iron is 0.07-0.1%, zinc is 0.05-0.1%, molybdenum is 0.03%, copper is 1.0%, and silver is 5g/t. Submeridional tectonic faults are of particular interest.

Conditional signs: Structural-decoding complexes: 1-Alluvial (sandy soil, loam, pebbles), 2-Proluvial (gravelites, pebbles, gravels),

3-Sandy-clay (clays, gravelites, sandstones, siltstones), 4-Molassic (clays, siltstones, sandstones), 5-Silicous-carbonate (limestones, marbles, marbled limestones with silicon), 6-Volcanogenic-terrigenous-shale (limestones, sandstones, siltstones, shales, tuffs), 7-Terrigenous-carbonate-shales (argillites, sandstones, limestones, siliceous), 8-Granitoid, 9-Olistostrome (mica-quartz shales, limestones, sandstones, dolomites), 10-ground cracks, 11-annular structures. **Elements of interpretation of geophysical fields of anomalous magnetic fields ΔTa, ΔZa:** 12-Certain positive ΔTa anomalies equated with distribution and pre-contact changes (hornification, skarning) of basic rocks, etc., 13- Axes of linear magnetic anomalies aligned with the zones where sulphide pyrite-pyrrhotite mineralization developed, 14- Longitudinal and transverse limiting main tectonic faults. **Residual anomalies of the gravitational field Δg:** 15-Contours of granitoid intrusives at depths of 100-1500 meters (stocks, apophyses or ceiling uplifts) expressed in rapidly negative Δg anomalies and positive Ta anomalies of variable dynamism, 16- Lineaments of the Δg field, uniformly leveled with fault faults between the blocks, set according to the descriptions of the gradient zones of the anomalies. 17-Regional tectonic faults separating megablocks. **ΔUTP is a natural potential field:** 18-Contours of areas of negative ΔUTP anomalies, uniformly equated with the distribution of strong cracks in the modified and sulphided rocks, 19-Zones of tectonic faults, ΔUTP anomaly axes of weak velocities uniformly leveled with contacts of rocks of different composition. **Types of bedrock according to the data of the broken wave comparison method:** 20-Outline contours of Tomezoic formations. **Mining and mineral manifestations:** 21-Gold ore manifestations. Mineralized points (according to the results of opening sampling): 22-Gold (gold content 0.1-1.0 g/t-calculated value) according to the results of previous work, 23-Gold (gold content 0.1-1.0 g/t-calculated value). Geochemical points (according to the results of opening sampling): 24-Gold amount 0.01-0.09 g/t, according to the results of previous work, 25-Gold, the amount of 0.01-0.09 g/t, according to the results of previous work, 26-Separate slags with a large amount of minerals: pure gold, 27-Tungsten, the amount of 0.0n-0.01%. 28-Arsenic, amount 0.01-0.02%, 29-Zinc-amount 0.01-0.2%, **Geochemical halos (distribution of secondary halos):** 30-Gold, 31-Mishchyak. Mud halos: 32-Pure gold, 33-Sheelet, 34-Skarnized rocks, 35-Monazite, 36-On the spread of contact metamorphism, 37-Longitudinal and transversal bounding main tectonic faults, 38-Sharyaj-thrust zones-tectonic boundaries of sedimentary complexes , 39- Potentially promising area.

Summary. The main features of deciphering geological and tectonic structures in space images are the texture and structure of the image, including relief, color (hue), shape of boundaries, shapes and sizes (geometric). As a result, these criteria made it possible to determine mineralization zones based on photoanomalies reflected in space images; As a result of the complex decoding of space images of the Southern Nurota area, it was possible to distinguish high-perspective ore positions. As a result, mineralization positions were confirmed; Field decoding observation works, visual and automatic decoding results were confirmed by the complex use of mineral-bearing areas. As a result, according to the results of field observation decoding in the region, mineral manifestations were observed at the intersection nodes of earth cracks in the north-eastern, sublatitude, and north-western directions; According to the results of cosmogeological studies, two prospective areas with high potential have been allocated in the Southern Nurota area. As a result, several positions adjacent to them were identified, and practical recommendations were given to relevant geological exploration enterprises for conducting preliminary geological exploration in the future.

CONCLUSION

1. The geological and tectonic structures of the South Nurata region, including granitoid, volcanogenic-carbonate-terrigenous, terrigenous, carbonate, clay-carbonate, silicon-carbonate, and mixtite structural-decoding complexes, have been identified using high-resolution satellite images from Landsat 7, Aster, and Quick Bird. These findings have been recommended for practical application due to their high potential.

2. The study has demonstrated the convenience of automatic decoding methods in studying the geological structures and tectonic features of the South Nurata region. These methods have significantly improved the quality indicators of geological decoding processes in high-potential satellite images. This advancement enhances the accuracy and reliability of geological investigations in the region.

3. Using remote sensing materials, the most effective RGB (Red, Green, Blue) channel combinations for solving geological problems in both open and closed areas have been identified. This has resulted in high efficiency in observing and automatically decoding geological objects.

4. The results of decoding digital space images were integrated with geological, geophysical, and geochemical data to create a large-scale remote-based cosmogeological map of the South Nurata region. This integration has led to the creation of an additional geological database for the studied area.

5. Comprehensive decoding of space images for the South Nurata region has led to the identification of highly promising mineral positions. As a result, two highly prospective areas and several adjacent positions have been identified and recommended to relevant geological exploration enterprises for preliminary geological exploration.

6. The remote-based cosmogeological map of the South Nurata region plays a vital role as a primary map for conducting regional analysis of endogenous mineralization and planning future geological exploration activities.

7. 7. The use of modern computer programs for decoding remote sensing materials, along with field control observations and reference objects, has created new opportunities for solving metallogeny issues in regional geological research. These advancements enhance the ability to interpret geological data accurately, aiding in the effective exploration and analysis of mineral resources.

LIST OF REFERENCES USED

Published by:

Абдуллаев Г.С., Долгополов Ф.Г. Геодинамика и нефтегазаносность литосферы Узбекистана. - Т.:Уз НИО НГП, 2016. -362 с.

Авезов А.Р. Методика применения материалов дистанционных съемок для выявления рудолокализующих структур в Западном Узбекистане (на примере Центральной части Нуратинских гор): Автореф.дисс. канд. геол.-мин.наук. -Т., 2004. - 22 с.

Азимов Б.Г. Применение аэрокосмических изображение в структурно-геологических исследованиях: Автореф.дисс. канд. геол.-мин.наук. - М., 1988. - 21с.

Асадов А.Р. Цифровая обработка данных дистанционного зондирования в решении структурных задач геологии в условиях Центральных Кызылкумов. // Мат-лы Республ. науч.-техн конф. «Приоритетные направления геологического изучения недр, гидрогеологических и инженерно-геологических исследований в Республики Узбекистан» -Т.:   ГП "НИИМР", 2011. - С. 17-19.

Асадов А.Р., Туляганова Н.Ш., Ахмадов Ш.И. Автоматизированное выделение линиаментов по радарным космоснимкам. // Мат-лы науч. конф. «Актуальные проблемы геологии, геофизики и металлогении». - Т.: ИГиГ им. Х.М. Абдуллаева, 2015. - С.286-289.

Атлас моделей рудных месторождений Узбекистана / Р.Х.Миркамалов и др. - Т: ГП «НИИМР», 2010. -100с.

Аэрокосмические и геолого-геофизические исследования закрытых платформенных территории /Трофимов Д.М., Богословский В.А., Ильина Е.Б. и др. - М.: Недра, 1986. - 238 с.

Баррет Э., Куртис Л. Введение в космическое землеведение. -М.: Прогресс, 1979. -230 с.

Борисов О.М. Глух А.К. Кольцевые структуры и линеаменты Средней Азии. - Т.: Фан, 1982. -122 с.

Бусыгин Б.С., Никулин С.Л., Бойко В.А. ГИС-технология поисков золота в Западном Узбекистане // Геоинформатика. - 2006. - №1. - С.44-49

Губин В.Н. Дистанционные методы в геологии: Учеб. пособие. -Минск БГУ, 2004. - 138 с.

Гоипов А.Б. Дешифровочные признаки структурно-вещественных комплексов на примере Южно-Ферганской территории // Тез. Межд. науч.-техн. конф. «Интеграция науки и практики как механизм эффективного развития геологической отрасли Республики Узбекистан». - Т.: ГП «ИМР», 2016. - С. 231-234.

Гоипов А.Б.Комплексирование геофизической, космогеологической информации для прогнозирования локальных нефтегазоперспективных структур в пределах Юго-Западных отрогов Гиссарского хребта: Автореф. дисс. док. философии геол-мин.наукам (PhD). - Т., 2019. - 47 с.

Гоипов А.Б., Усманов М.С., Тогаев И.С. Спектральные отражательные способности горных пород на примере гор Актау (Южный Нуратау). // Мат-лы Междунар. науч.-техн. конф. «Интеграция науки и практики как механизм эффективного развития геологической отрасли Республики Узбекистан». - Т.: ГП «ИМР», 2018, - С.171-173

Геологический словарь. В трех томах. Издание третье, перераб, и доп. / Гл. ред. О.В.Петров. Т.1.-СПб.: ВСЕГЕИ, 2012. - 432 с.

Геологический словарь. В трех томах. Издание третье, перераб, и доп. / Гл. ред.О.В.Петров. Т.2. –СПб.: ВСЕГЕИ, 2012. -480 с.

Глух А.К., Авезов А.Р. Некоторые аспекты геодинамики и возраста кольцевых структур // Геология и минеральные ресурсы. -2005. -№ 3. -С.8-11.

Глух А.К., Поторжинский М.Г., Эйсфельд О.А. Автоматизированное структурное дешифрирование материалов космических съемок // Геология минеральных ресурсов. - 2009. -№2. -С.35-41.

Глух А.К., Мехмонходжаев А.Д., Ким К.Н. Компьютерные технологии – путь к новым методам интерпретации космических снимков // Геология и минеральные ресурсы. -2003. -№4. -С.4-7.

Гонсалес Р. Мир цифровой обработки. Техносфера. - М., 2006. -1072 с.

Загубный Д.Г. Способы обработки цифрового рельефа программой «Lineament» // Исследование Земли из космоса. - 2004. - № 6 - С.30-38.

Златопольский А.А., Малкин Б.В. Автоматизированный анализ ориентационных характеристик данных дистанционного зондирования (программа «LESSA») // Современные проблемы ДЗЗ из космоса. -2004. - Т. 2. - С.188-195.

Кашкин В.Б., Сухинин А.И. Дистанционное зондирование земли из Космоса. Цифровая обработка изображений: Учебное пособое. - М.: Логос, 2001. - 260 с.

Кац Я.Г., Рябухин А.Г., Трафимов Д.И. Космические методы в геологии. - М.: Изд-во МГУ, 1976. -248с.

Кац Я.Г., Полетаев А.И., Румянцева Э.Ф. Основы линеаментной тектоники. -М.: Недра, 1986.-140с.

Кац Я.Г., Тевелев А.В., Полетаев А.И. / Основы космической геологии. Учебное пособое. - М.: Недра, 1988, -235 с.

Кронберг П. Дистанционное изучение Земли. Основы и методы дистанционных исследований в геологии. - М.: Мир, 1998. - 343 с.

Крутиков А.Д. Автоматизированное комплексирование геологических, геофизических данных и материалов дистанционного зондирования в ГИС // Мат-лы Междунар. науч.-тех.конф. «Современные проблемы гидрогеологии, инженерной геологии, геоэкологии и пути их решения». - Т.: ГП «Институт ГИДРОИНГЕО», 2015. - С. 317-319.

Крутиков А.Д. Алгоритмы автоматизированного выявления геологических линейных и кольцевых структур, по материалам дистанционного зондирования // Там же. -С. 320-322.

Миркамалов Р.Х. Этапы структурной перестройки домезозойского складчатого основания Нуратинского горнорудного района // Геология и минеральные ресурсы. -2008. - №6. -С. 6-16.

Миркамалов Р.Х., Диваев Ф.К., Селтман Р., Конопелько Д.Л. Геодинамическая эволюция магматизма и связанного с ним оруденения Западного Тянь-Шаня на территории Республики Узбекистан // Геология и минеральные ресурсы. -2018.- №1. - С.3-15.

Нурходжаев А.К. Современное состояние и перспективы развития дистанционного зондирования Земли в области геологоразведочного производства Республики Узбекистан // Геология и минеральные ресурсы. - 2017. - №1. - С. 78-83.

Нурходжаев А.К., Асадов А.Р., Рискидинов Ж.Т., Убайдуллаева Ш.А. Космоструктурные особенности гор Тамдытау и некоторые результаты космогеологических исследований // Геология и минеральные ресурсы. -2015. - №4. - С.9-12.

Нурходжаев А.К., Тогаев И.С., Шамсиев Р.З. Методическое руководство по составлению космогеологической карты Республики Узбекистан на основе цифровых космоснимков. -Т.: ГП «ИМР», 2017. -200с.

Нурходжаев А.К., Тогаев И.С., Шамсиев Р.З. Дистанционная основа Республики Узбекистан: создание, требование и перспективы ее использования при планировании региональных геологических исследований // Геология и минеральные ресурсы. - 2017. - №4. - С.3-6.

Нурходжаев А.К., Тогаев И.С. Некоторые результаты геологических исследований по выявлению потенциально прогнозных участков на основе данных дистанционного зондирования Земли (напримере пл. Шаркий Каулюк) // Геология и минеральные ресурсы. -2018. -№2.-С.3-10.

Нурходжаев А.К., Тогаев И.С., Ибрагимов Р.Х., Хасанов Н.Р. Инновационная технология дешифрирования материалов дистанционного зондирования земли для решения прикладных задач геологии // Геология и минеральные ресурсы. - 2019. - №3. - С.3-8.

Нурходжаев А.К. Структурно-геодинамические и геологические основы дистанционного зондирования Земли в целях решения прогнозно-поисковых задач (Западная часть Тянь-Шаня). Автореф. дисс. докт. г.-м. наук. -Т., -2019. - 57 с.

Нурходжаев А.К., Тогаев И.С. К вопросу создания дистанционных основ по структурно-дешифровочным комплексам // Мат-лы науч.-техн. конф., посвящ. 80-летию создания института геологии и геофизики и 105-летию со дня рождения акад. Х.М.Абдуллаева. - Т.: ИГиГ им.Х.М. Абдуллаева, 2017. - С.258-259.

Нурходжаев А.К., Тогаев И.С. Новый методический подход к технологии дешифрирования материалов дистанционного зондирования Земли // Мат-лы I Междунар. науч.-техн. конф. «Роль науки и практики в усилении устойчивости и актуализации управления рисками проявления экзогенных геологических процессов». - Т., 2019. - С.132-134.

Перцов А.В., Антипов В.С. Наукоемкие технологии космо-аэрогеологических исследований для обеспечения современного уровня ведения металлогенических, прогнозных и поисковых работ // Руды и металлы. - 2002. - №3. - С. 80-81.

Пирназаров Т.П., Тогаев И.С., Соостер А.Е. Геологическое строение и космогеологические особенности размещения золотого орудения на Шаркий Каулюкской перспективной пл. Мадмонского рудного поля // Мат-лы Междунар. науч.-техн. конф. -Т.: ГП «НИИМР», 2016. - С.314-317.

Пирназаров М.М., Асадов А.Р., Пирназаров М.М. Космоструктурно-геохимический метод прогнозирования золото-редкометалльного оруденения в горах и предгорных районах Узбекистана // Геология и минеральные ресурсы. - 2017. -№6. - С. 64-68.

Пятков А.К. Тектоника западного окончания гор Нуратау и прилегающих территорий. Геология и минеральные ресурсы. - 2000, -№2. -С.16-20.

Рудные месторождения Узбекистана / Под ред. И.М.Голованова. - Т.: ГИДРОИНГЕО, 2001. - 660 с.

Сидорова И.П. Строение литосферы Кызылкумского и Нуратинского рудных регионов Узбекистана // Мат-лы Междунар. науч.-техн. конф. «Интеграция науки и практики как механизм эффективного развития геологической отрасли Республики Узбекистан». -Т.: ГП «ИМР», 2018. - С. 100-104.

Снитка Н.П., Хамроев И.О. Связь геодинамики с металлогенией Южного Тянь-Шаня и актуальные задачи ведения геолого-разведочных работ // Мат-лы науч. конф., посвящ. 80-летию акад. Т.Н.Далимова. «Геодинамика, магматизм и оруденение Западного Тянь-Шаня». - Т., 2016. - С. 105-109.

Стратифицированные и интрузивные образования Узбекистана. -Т.: ИМР, 2000. -511 с.

Токарева О.С. Обработка и интерпретация данных дистанционного зондирования Земли: учебное пособие. - Томск: Изд-во Томского политехн. ун-та, 2010. -148 с.

Тогаев И.С., Нурходжаев А.К. Структурно-дешифрируемые комплексы выделяемые на основе дистанционного зондирования Земли территории Узбекистана // Вестник НУУз. - 2019. - №3/1. - С. 341-347.

Тогаев И.С., Нурходжаев А.К. Некоторые результаты комплексного исследования территории гор Южный Нуратау // Геология и минеральные ресурсы. -2019. - №5. - С.3-7.

Тоғаев И.С. Замонавий усуллар ва технологиялар ёрдамида регионал тадқиқотларнинг айрим масалалари (Жанубий Нурота мисолида) // Геология ва минерал ресурслар. -2020. - №1. - С.90-94.

Тогаев И.С., Рузиев С.К., Хасанов Н.Р., Гоипов А.Б. Структурно-дешифрируемые комплексы магматических интрузивных образований Узбекистана, выделяемые на основе дистанционного зондирования Земли // Мат-лы Междунар.науч.-техн. конф. «Интеграция науки и практики как механизм эффективного развития геологической отрасли Республики Узбекистан». - Т.: ГП «ИМР», 2018. - С. 113-115.

Тогаев И.С., Нурходжаев А.К. Дистанционная основа Южного склона гор Актау (Южный Нуратау). // Мат-лы I Междунар. науч.-техн. конф. «Роль науки и практики в усилении устойчивости и актуализации управления рисками проявления экзогенных геологических процессов». - Т., 2019. - С. 158-160.

Тогаев И.С., Нурходжаев А.К. Дистанционная основа гор Южный Нуратау и ее роль в региональных исследованиях // Междунар. науч.-практ. конф. «Актуальные проблемы нефтегазовой геологии и инновационные методы и технологии освоения углеводородного потенциала недр» Т.: АО «ИГиРНиГМ», 2019. - С. 420-423.

Тогаев И.С., Нурходжаев А.К. Жанубий Нурота тоғларида олиб борилган космогеологик изланишлар ва уларнинг айрим натижалари // Халқаро ёш олимлар анжумани. «Фан ва инновация». - Т., 2019. - С. 196-198.

Тогаев И.С., Нурходжаев А.К. Геологическая информативность материалов дистанционного зондирования Земли при космогеологических исследованиях (на примере гор Южный Нуратау) // Тез.докл. I Молодежной науч.-образов. конф. ЦНИГРИ. –М.: ФГБУ «ЦНИГРИ», 2020 - С. 189-191.

Тогаев И.С., Нурходжаев А.К. Космогеологик изланишлар ва уларни минтақавий тадқиқотларда аҳамияти (Жанубий Нурота мисолида) // ЎзМУ Геология ва геоинформацион тизимларнинг долзарб муаммолари. Республика илмий ва илмий-техник анжуман. - Т., 2020. - Б.64-68.

Тоғаев И.С. Нурходжаев А.К. Космогеологик тадқиқотларда замонавий усуллар (Жанубий Нурота мисолида). // ТДТУ. Республикада геология ўқитишининг долзарб муаммолари ва Ер фанлари истиқболлари. Республика илмий ва илмий-техник анжуман мат-лари. - Т., 2020. - Б.135-138.

Феодоров Е.Г., Миркамалов Р.Х., Диваев Ф.К. Олистостромовые толщи Западного Тянь-Шаня // Геология и минеральные ресурсы. -2016. - №3. - С. 3-12.

Чандра А.М. Дистанционное зондирование и географические информационные системы / Отв. ред. С.К.Гош. -М.: Техносфера, 2008. -312 с.

Черново И.Ю., Нугманов И.И., Кадыров Р.И. Учебно-методическое пособие «Автоматизированный линеаментной анализ». -Казань, 2012. - 38с.

Шамсиев Р.З. Алгоритмы автоматизированного выделения потенциальных рудных полей с использованием известных эталонных объектов // Мат-лы Междунар. науч.-тех.конф. «Современные проблемы гидрогеологии, инженерной геологии, геоэкологии и пути их решения». - Т.: ГП «Институт ГИДРОИНГЕО», 2015. - С.343-345.

Эргашев Ш.Э., Асадов А.Р. Методические рекомендации по использованию дистанционных съемок. - Т.: ИМР, 2001. - 224 с.

Эргашев Ш.Э., Исаходжаев Б.А. Асадов А.Р. Цифровые данные теледетекции в изучении вещественного состава горных пород // Геология и минеральные ресурсы. - 2007. -№2. - С. 6-11.

Эргашев Ш.Э., Закиров О.Т. Современные методы поиска рудной минерализации в условиях низкогорья. -Т.: ГП «ИМР», 2009. -119 с.

Эргашев Ш.Э., Шульц С.С. и др. Разработка поисковых признаков золотого оруденения на основе аэро- и космических снимков в полузакрытых условиях Центральных Кызылкумов. -Т.: САИГИМС, 1985.

Якимов А.А., Крутиков А.Д. Применение данных тепловых каналов ночных КС Aster в геологии. // Мат-лы Междунар. науч.-тех.конф. «Современные проблемы гидрогеологии, инженерной геологии, геоэкологии и пути их решения». - Т.: ГП «Институт ГИДРОИНГЕО», 2015. - С.-350-352.

Alkan M., Oruc M., Yildirim Y., Seker D. Z. Monitoring Spatial and Temporal Land Use/Cover Changes; a Case Study in Western Black Sea Region of Turkey // Journal of the Indian Society of Remote Sensing 2013. - № 41(3). -P.587-596.

Busygin B., Inozemtsev S., Nikulin S. The integrated analysis of geologicalgeophysical and remote sensing data at the gold prospecting in Western Uzbekistan // 67rd EAGE conference: Extended Abstracts. Madrid, Spain, 2005. - P. 346

Goipov and others. Structural features of the South-Western Gissar region according to Earth remote sensing data. International Journal of Geology, Earth and Environment Science. - 2019.-Vol. 9. (1). - P. 45-52.

Nurhodjaev A.K., Remote methods for solving the applied problems of geology in the light of modern requirements // International Journal of Geology, Earth and Environment Science. - India. 2018.-Vol. 8. - №1.- P. 1-4.

Nurhodjaev A.K., I.S.Togaev, R.X.Ibragimov. Materials of Remote Sensing of the earth for solving the applied problems of geology // European Applied Sciences. - Germany. -2018. -№.3. - P. 28-30.

Sarma J.N., Acharjee Sh Application of DEM, Remote Sensing and Geomorphic Studies in Identifying a Recent [or perhaps Neogene?] Upwarp in the Dibru River Basin, Assam, India // Journal of the Indian Society of Remote Sensihg. - 2011.- №39 (4). - P. 507-515.

Togaev I.S., Nurkhodjaev A.K., Akmalov Sh. Structurally decryptable complexes – a new taxonomic unit in cosmo-geological research // Journal E3S Web of conferences. - 2020. V.2. - №63. - P.1-8.

Togaev I.S. On the question of creating remote bases on materials of remote sensing of the earth on the example of the territory of the mountains of southern Nuratau // International Journal of Geology, Earth & Environmental Sciences. -2020 Vol.10 (2). - P.34-40.

Regarding the fund:

Глух А.К. Совершенствование методик обработки результатов дешифрирования материалов космических съемок с целью повышения эффективности геологоразведочных работ координация аэрокосмогеологических исследований в организациях Госкомгеологии. Отчет. - Т., 2009. -193 с.

Колоскова С.М. и др. Выделение перспективных на золото поисковых площадей в Южном Нуратау на основе комплекса геолого-геохимических критериев. Отчет за 2001-2005 гг. - Т.:ИМР, 2005.

Нурходжаев А.К., Тогаев И.С. Составление дистанционных основ – космогеологических карты в масштабе 1:500 000 по территории Республики Узбекистан. - Т.: ГГФ, 2016.

Пяновская И.А. и др. Составление аэрофотогеологической карты гор Южные Нуратау и прилагающих территорий масштаба 1:50000 (по работам 1980-1984гг). -Ташкент-Самарканд: ГГФ, 1984.

Рискидинов Ж.Т. Мовланов Ж.Ж. Изучение условий размещения золотого и вольфрамового оруденения в западной части Южно-Нуратинских гор с использованием новых цифровых м-лов теледетекции.-Т.:ИМР, 2012.

Тогаев И.С. «Опережающие космогеологические исследования в горах Южный Нуратау» за 2015-2018гг. - Т.: ГГФ, 2018.

Хачатурян Е.Р. Авезов А.Р., Глух А.К. Космогеологические исследования среднего (1:200 000) и крупного (1:50 000) масштабов территории Узбекистана с использованием материалов дистанционного зондирования Земли». - Т.: ГГФ, 2011.

Хан Р.С. Геологические строение и полезные ископаемые гор Актау и Каракчатау. Отчет Актауской ГСП о результатах доизучения (ГДП-2) складчатого комплекса с общими поисками на золото, ртуть и др. полезных ископаемых гор Актау и Каракчатау масштаба 1:50000. Зарафшанская ГПЭ. - Самарканд: ГГФ, 1989.

Хан Р.С. Геологические доизучение на площади 1572 кв.км. и геологическая съемка (ГС) на площади 98 кв.км. в центральной части хребта Южный Нуратау масштаба 1:50000. Зарафшанская ГПЭ. -Самарканд: ГГФ, 2002.

<h1 style="text-align:center">TABLE OF CONTENTS</h1>

Printed by Libri Plureos GmbH in Hamburg,
Germany